Typesetting Documents in Scientific WorkPlace® and Scientific Word®

Second Edition

A Guide to Typesetting with Scientific WorkPlace and Scientific Word

Typesetting Documents in Scientific WorkPlace® and Scientific Word®

Second Edition

A Guide to Typesetting with Scientific WorkPlace and Scientific Word

Susan Bagby
George Pearson
MacKichan Software, Inc.

©2003 by MacKichan Software, Inc. All rights reserved. No part of this book may be reproduced, stored in a retrieval system, or transcribed, in any form or by any means—electronic, mechanical, photocopying, recording, or otherwise—without the prior written permission of the publisher, MacKichan Software, Inc., Poulsbo, Washington. Information in this document is subject to change without notice and does not represent a commitment on the part of the publisher. The software described in this document is furnished under a license agreement and may be used or copied only in accordance with the terms of the agreement. It is against the law to copy the software on any medium except as specifically allowed in the agreement.

Printed in the United States of America

10 9 8 7 6 5 4 3 2 1

Trademarks

Scientific Word, *Scientific WorkPlace*, *Scientific Notebook*, and EasyMath are registered trademarks of MacKichan Software, Inc. EasyMath is the sophisticated parsing and translating system included in *Scientific Word*, *Scientific WorkPlace*, and *Scientific Notebook* that allows the user to work in standard mathematical notation, request computations from the underlying computational system (MuPAD in this version) based on the implied commands embedded in the mathematical syntax or via menu, and receive the response in typeset standard notation or graphic form in the current document. MuPAD is a registered trademark of SciFace GmbH. Acrobat is the registered trademark of Adobe Systems, Inc. T$_E$X is a trademark of the American Mathematical Society. TrueT$_E$X is a registered trademark of Richard J. Kinch. PDFT$_E$X is the copyright of Hàn Thế Thành and is available under the GNU public license. Windows is a registered trademark of Microsoft Corporation. MathType is a trademark of Design Science, Inc. All other brand and product names are trademarks of their respective companies. The spelling portion of this product utilizes the Proximity Linguistic Technology. Words are checked against one or more of the following Proximity Linguibase® products:

Linguibase Name	Publisher	Number of Words	Proximity Copyright
American English	Merriam-Webster, Inc.	144,000	1997
British English	William Collins Sons & Co. Ltd.	80,000	1997
Catalan	Lluis de Yzaguirre i Maura	484,000	1993
Danish	IDE a.s	169,000	1990
Dutch	Van Dale Lexicografie bv	223,000	1996
Finnish	IDE a.s	191,000	1991
French	Hachette	288,909	1997
French Canadian	Hachette	288,909	1997
German	Bertelsmann Lexikon Verlag	500,000	1999
German (Swiss)	Bertelsmann Lexikon Verlag	500,000	1999
Italian	William Collins Sons & Co. Ltd.	185,000	1997
Norwegian (Bokmal)	IDE a.s	150,000	1990
Norwegian (Nynorsk)	IDE a.s	145,000	1992
Polish	MorphoLogic, Inc.		1997
Portuguese (Brazilian)	William Collins Sons & Co. Ltd.	210,000	1990
Portuguese (Continental)	William Collins Sons & Co. Ltd.	218,000	1990
Russian	Russicon		1997
Spanish	William Collins Sons & Co. Ltd.	215,000	1997
Swedish	IDE a.s	900,000	1990

This document was produced with *Scientific WorkPlace*.

Authors: *Susan Bagby and George Pearson*
Editorial Assistant: *John MacKendrick*
Compositor: *MacKichan Software Inc.*
Printing and Binding: *Malloy Lithographing, Inc.*

Contents

What's New vii
    About This Manual vii
    Typesetting Basics viii
    Conventions xi
    Getting Help xiii

1 Tailoring Typesetting to Your Needs 1
    Cautions 1

    Tailoring the Page Layout 2
    Changing the Margins 2
    Changing the Line Spacing 3
    Changing the Font Size of Body Text 5
    Changing the Headers and Footers 5
    Changing the Page Numbering 9
    Creating Multiple Columns 12
    Changing the Page Orientation 13
    Changing to a Different Paper Size 14

    Tailoring the Front and Back Matter 16
    Changing the Title Page 16
    Changing the Table of Contents 17
    Creating an Unnumbered Manual Bibliography 20
    Creating an Appendix for Each Chapter 21
    Creating Two Separate Indices 21

    Tailoring the Body of the Document 23
    Changing the Appearance of Division Headings 23
    Changing the Appearance of Paragraphs 27
    Changing the Appearance of Numbered Lists 28

vi   Contents

    Underlining and Striking Through Content   30
    Stopping Hyphenation   31
    Changing Footnotes to Endnotes   31
    Using Color in Documents   32

  Tailoring Theorem Environments   33
    Changing Theorem Numbering   33
    Changing Theorem Formatting   35

  Tailoring Graphics and Tables   36
    Formatting Graphics and Table Captions   36
    Wrapping Text Around Graphics   38
    Managing EPS Graphics   40
    Managing Floating Objects   41
    Creating Landscaped Tables   42

  Making Additional Typesetting Changes   42

  Troubleshooting   48
    Resolving LaTeX Errors   49
    Repairing a Damaged Document   50

2   Using Document Shells   53

  Understanding Document Shells   53
    LaTeX Document Classes   54
    LaTeX Class Options   55
    LaTeX Packages   60

  Choosing a Document Shell   61

  Tailoring a Shell to Your Needs   63
    Modifying the Document Class Options   64
    Adding and Removing LaTeX Packages   65
    Modifying LaTeX Package Options   67
    Adding TeX and LaTeX Commands to a Document   68

  Creating Shells   70

  Using Shells and Typesetting Specifications from Outside Sources   71

3   Using LaTeX Packages   77

Index   149

What's New

Typesetting Documents is a response to the many users who have requested more information about LaTeX typesetting in *Scientific WorkPlace (SWP)* and *Scientific Word (SW)*. This new edition contains information about typesetting in Versions 3.5, 4, and 5 of *SWP* and *SW*.

About This Manual

Our purpose is threefold:

- to answer common questions about typesetting in *SWP* and *SW;*
- to help you choose document shells appropriately;
- to explain when and how you can tailor document shells *from within the program* so that you can create documents that more precisely meet your typesetting needs.

To these ends, we begin the manual with what may be the only thing you need to read: the answers to the questions users ask most frequently about LaTeX typesetting in *SWP* and *SW*. Chapter 1 "Tailoring Typesetting to Your Needs" provides tips on modifying page layout, document elements, tables and figures, and mathematics in *SWP* and *SW* documents. It also offers troubleshooting information.

If you need more basic information, we suggest you start with Chapter 2 "Using Document Shells," which discusses the most basic and perhaps most important document development task: how to choose a document shell—or template—that meets the typesetting requirements of you and your publisher. The chapter discusses how and to what extent you can tailor a document using available document class and LaTeX package options and, where necessary, TeX commands. Chapter 2 also explains how to create *SWP* and *SW* documents using shells obtained from an outside source.

Chapter 3 "Using LaTeX Packages" focuses on LaTeX packages provided with the program. The discussion notes how each package interacts and, occasionally, conflicts with other typesetting elements.

Accompanying this manual is another volume, *A Gallery of Document Shells for Scientific WorkPlace and Scientific Word,* available on your program CD as a PDF file. *A Gallery of Document Shells* explains the key characteristics of the shells provided with the program. It contains brief discussions and illustrations of typeset documents created with each shell except those created with *Scientific Notebook*, which were intended for direct printing instead of typesetting.

Please note that this manual doesn't apply to *Scientific Notebook* or *Scientific Viewer*, which don't support typesetting. The manual excludes a discussion of the processes involved in producing a document without typesetting. Further, it excludes an in-depth discussion of working with document shells created with the Style Editor. Extensive Style Editor documentation is available online (see Online Help on page xiii).

This manual assumes that you have successfully installed *SWP* or *SW* and that you have a working knowledge of the program. Although certain procedures are explained here, you will find fuller explanations in these accompanying manuals:

- *Getting Started with Scientific WorkPlace, Scientific Word, and Scientific Notebook*
- *Creating Documents with Scientific WorkPlace and Scientific Word*
- *Doing Mathematics with Scientific WorkPlace and Scientific Notebook*

The instructions in this manual occasionally differ for different versions of *SWP* and *SW*. When several sets of instructions are provided, be sure to choose the correct set of instructions for your version of the product.

This manual also assumes you're familiar with basic TEX, the extraordinary mathematics typesetting program and language designed by Donald Knuth, and with LATEX, the set of macros designed by Leslie Lamport to enhance TEX with document-structuring features such as tables of contents, chapters and sections, lists, and bibliographies. The Windows implementation of TEX and LATEX that is supplied with the program is TrueTEX, a product of TrueTEX Software. The TrueTEX software distributed with Version 5 of *SWP* and *SW* includes PDFTEX support.

The manual often includes detailed information about TEX and LATEX, document classes, LATEX packages, typesetting options, and document shells. While a good understanding of TEX and LATEX will help you better understand how these often complex elements interact, a thorough discussion of TEX and LATEX is beyond our scope here. If you need additional information, we suggest you refer to these excellent sources:

- *The TEXbook* by Donald E. Knuth
- *LATEX, A Document Preparation System* by Leslie Lamport
- *The LATEX Companion* by Michel Goossens, Frank Mittelbach, and Alexander Samarin
- *A Guide to LATEX: Document Preparation for Beginners and Advanced Users* by Helmut Kopka and Patrick W. Daly
- The TEX Users Group website at http://www.tug.org
- The Usenet newsgroup news:comp.text.tex

Typesetting Basics

With Version 3.0 and later of *SWP* and *SW*, you can produce your documents either with or without LATEX typesetting. Our focus here is on typesetting. Even if you have a basic understanding of the program, it's important to review how typesetting works and why the appearance of your printed document differs so noticeably when you typeset compared to when you don't. You can find basic information in more detail in the online Help and in *Creating Documents with Scientific WorkPlace and Scientific Word*.

With Version 5, we introduce an important new feature: creating typeset Portable Document Format (PDF) files with PDFTEX, which provides all the beauty and features of LATEX typesetting in PDF form. Now you can use *SWP* and *SW* to typeset files for viewing across platforms with PDF viewers.

Throughout our documentation, we refer to the processes that you use to typeset your documents with LATEX as *typeset compile, typeset preview,* and *typeset print.* We refer to the processes that you use to typeset your documents with PDF as typeset *compile*

PDF, typeset preview *PDF,* and typeset *print PDF.* All these processes are available as commands on the **Typeset** menu or as buttons on the Typeset toolbar. In general, we use the term *typesetting* to refer to either set of processes. Where necessary, we distinguish between them. We refer to the processes that don't involve typesetting as *preview* and *print.* These commands are available on the **File** menu and the Standard toolbar.

Producing Documents with Typesetting

When you process your document with LaTeX, the program compiles it with LaTeX to create a *device independent*—or *DVI—file,* which is a finely typeset version of your document. The DVI file may contain automatically generated document elements such as cross-references, tables of contents, and numbers for equations. The program then sends the DVI file to the typeset previewer or to the printer.

Typesetting with PDFTeX in Version 5 is just like typesetting with LaTeX, except that you produce a PDF file. Like the DVI file, the PDF file is a finely typeset version of your document. It contains the same automatically generated document elements as the DVI file and, if you have added the *hyperref* package to your document, live hypertext links. The PDF file also contains embedded fonts and, if you have specified PDF output settings, graphics that have been converted to formats appropriate for PDF viewers. The program sends the PDF file to the screen or the printer using your PDF software.

The appearance of the PDF and DVI files is almost identical. However, a typeset document has a noticeably different appearance from what you see as you work in the document window or when you produce your document without typesetting.

The typeset appearance of your document depends on typesetting specifications from three different sources, all set initially by the shell you use to create your document:

- The *typesetting specifications,* a collection of TeX and LaTeX instructions related to typesetting document elements, including those represented by the tags on the Tag toolbar in the program window
- Any LaTeX *packages* or *options* specified for the document or the shell
- Any additional LaTeX *commands* that appear in the document preamble or in the body of the document

These specifications don't affect the appearance of your document if you don't typeset.

The chapters that follow explain how to modify some of these specifications from within *SWP* and *SW.* However, we advise against attempts at extensive modification of the specifications if you aren't extremely familiar with TeX and LaTeX.

▶ **To typeset a document**

1. If you have Version 5, select an output option for typesetting:
   a. From the **Typeset** menu, choose **Output Choice**.
   b. Select the option you want. **DVI Output** is the default.
   c. If you're creating a PDF file,
      i Check **Convert .tex link targets to .pdf** to convert `.tex` extensions in hypertext link addresses to `.pdf` extensions.
      ii Choose **PDF Graphics Settings** to specify the file type and location of graphics and plots exported during the typesetting process, and then choose **OK**.
   d. Choose **OK**.

2. On the Typeset toolbar, click the button for the typesetting command you want or, from the **Typeset** menu, choose the command (commands for typesetting with PDFLaTeX appear only in Version 5):

Menu	Command	Button		Menu	Command	Button
Typeset	Compile			Typeset	Compile PDF	
Typeset	Preview			Typeset	Preview PDF	
Typeset	Print			Typeset	Print PDF	

3. If you choose a **Compile** command, select the options you want and choose **OK**.

    Note that the **Compile** dialog box indicates the name and location of the typeset file to be created.

4. If you choose a **Preview** or **Print** command, set the number of LaTeX passes.

    The program compiles your document if necessary. When the compilation is complete, the program either displays your typeset document in the TrueTeX Previewer or prints it on the printer you indicate. If you chose to create a typeset PDF file, the program opens your PDF viewer to display the typeset file or print your document.

You can also print your document from the TrueTeX Previewer or your PDF viewer.

Producing Documents without Typesetting

When you produce your document without typesetting it, the program sends the document directly to a non-LaTeX previewer or to the printer using many of the same routines with which it displays the document in the document window. Consequently, what you see in the preview window or in print is similar to what you see as you work on your document in the document window. (The program doesn't reflect the page setup specifications or the print options in the document window.)

When you don't typeset, the appearance of your document depends on three sets of specifications; like the typesetting specifications, they are all set initially by the document shell:

1. The *style,* a collection of the specifications for the appearance of each tag in the document window and in print.

2. The *page setup specifications.*

3. The *print options.*

The online Help and *Creating Documents with Scientific WorkPlace and Scientific Word* provide information about modifying these specifications and about previewing and printing without typesetting. These three sets of specifications don't affect the typeset appearance of your document in any way, although the style determines how the document appears on the screen when you're working on it.

Understanding the Differences in the Final Product

Each time you produce your document in *SWP* or *SW*, you can choose whether or not to typeset it. The results differ noticeably.

If you typeset, the program compiles the document and generates any specified automatic elements such as front matter items (tables of contents or lists of figures and tables), cross-references, footnotes and margin notes, automatically numbered equations, indexes, and bibliographies. LaTeX and PDFLaTeX also provide hyphenation, kerning, ligatures, sophisticated paragraph and line breaking, and other automatic formatting features.

If you don't typeset, the program produces the document using many of the same routines it uses to display the document in the document window. No document elements are automatically generated, and the printed results are similar to what you see as you work on the document.

Conventions

Understanding the notation and the terms used in our documentation will help you understand the manual instructions. We assume you're familiar with basic Windows procedures and terminology. If necessary, review your Windows documentation. In our manuals, we use the notation and terms listed below.

General Notation

- **Text like this** indicates information you should type exactly as it is shown.
- Text like this indicates the name of a menu, command, or dialog.
- TEXT LIKE THIS indicates the name of a keyboard key.
- `Text like this` indicates the name of a file or directory.
- ***Text like this*** is a placeholder for a file name or other information that you must supply.
- *Text like this* indicates a term that has special meaning in the program.
- The phrase *typeset your file* means to process your document with either LaTeX or, if you have Version 5, PDFLaTeX.
- The word *choose* means to designate a command for the program to carry out. As with all Windows applications, you can choose a command with the mouse or with the keyboard. Commands may be listed on a menu or shown on a button in a dialog box. For example, the instruction "From the File menu, choose Open" means you should first choose the File menu and then from that menu, choose the Open command. The instruction "choose OK" means to click the OK button with the mouse or press TAB to move the attention to the OK button and then press the ENTER key on the keyboard.
- The word *select* means to highlight the part of the document that you want your next action to affect or to highlight a specific option in a dialog box or list.
- The word *check* means to turn on an option in a dialog box.

Keyboard Conventions

We also use standard Windows conventions to give keyboard instructions.

- The names of keys in the instructions match the names shown on most keyboards. They appear like this: ENTER, F4, SHIFT.
- A plus sign (+) between the names of two keys indicates that you must press the first key and hold it down while you press the second key. For example, CTRL+G means that you press and hold down the CTRL key, press G, and then release both keys.
- The notation CTRL+**word** means that you must hold down the CTRL key, type the word that appears in bold type after the +, then release the CTRL key. Note that if a letter appears capitalized, you should type that letter as a capital.

Mouse Conventions

The program uses these mouse pointers:

Pointer	Indication
I	The pointer is over text
▶	The pointer is over mathematics
▶	A selection is being dragged
▶	A selection is being copied
▶	A selection is being copied or dragged with the right mouse button
✋	A graphic is being panned
+	A graphic is being resized
☝	The pointer is over a hypertext link

Additionally, the program displays a pointer for the computational engine when a computation is in progress.

In this manual we give mouse instructions using standard Windows conventions. The instructions assume you have not changed the mouse button defaults.

- *Point* means to move the mouse pointer to a specific position.
- *Click* means to position the mouse pointer, then press and immediately release the left or right mouse button without moving the mouse.
- *Double-click* means to position the mouse pointer, then click the left mouse button twice in rapid succession without moving the mouse.
- *Drag* means to position the mouse pointer, press the left mouse button and hold it down while you move the mouse to a new location, then release the button.

As in most Windows applications, you can use the right mouse button to display a Context menu for the current selection or the item under the mouse pointer. Pressing the Application key also displays the menu.

Getting Help

In addition to the information available in the manuals supplied with the program, you can get information about *SWP* and *SW* from the online Help system, the library of reference materials about mathematics and science, and, if you have an Internet connection, the MacKichan Software website. If these resources don't contain the information you need, technical support is available. We also regularly make additional information available on our unmoderated discussion forum and electronic mail list. You can find an errata sheet for this book, as well as all other manuals published by MacKichan Software, Inc., at this URL: **http://www.mackichan.com/techtalk/errata.html**.

Online Help

Without leaving *SWP* and *SW*, you can search the online Help system to find basic and advanced information about all program commands and operations, including those related to numeric, symbolic, and graphic computations. In particular, you can find additional material regarding T_EX, $L\!A\!T_EX$, $L\!A\!T_EX$ packages, and other related topics. If you save copies of the Help documents in *SWP*, you can interact with the mathematics they contain, experimenting with or reworking the included examples. In addition, two associated programs—the Style Editor and the Document Manager—have their own online Help systems.

▶ **To get help from the Help menu**

Choose	To
Contents	See a list of online information
Search...	Find a Help topic
Index	Access the online index to Computing Techniques, General Information, or the Reference Library
MacKichan Software Website	Open the link to the MacKichan Software, Inc. website
Register...	Register your software and obtain a license
System Features...	See a list of available features; change the serial number for your installation
License Information	Obtain information about registering your system
About...	Obtain information about your installation

▶ **To go directly to the Help Contents, press** F1.

Supplemental Technical Documents

We urge you to explore the supplemental technical documents supplied with the program. You can use *SWP* or *SW* to open, view, and print the documents. In particular, we urge you to read the following documents:

- In the `Help\general` directory, the document `50techref.tex`, which contains technical information on the features in Version 5.
- In the `Play` directory, the sample documents, which demonstrate the use of computation in *SWP*.
- In the `SWSamples` directory,
    - The sample documents, which illustrate the use of various LaTeX packages in *SWP* and *SW*.
    - The file `OptionsPackagesLaTeX.tex`, which describes and contains links to information about the options, packages, and other TeX-related items provided with the program.
    - The file `BibTeXBibliographyStyles.tex`, which lists and describes the BibTeX style (`.bst`) files installed with the program.

Obtaining Technical Support

If you can't find the answer to your questions in the manuals or the online Help, you can obtain technical support from our website at

http://www.mackichan.com/techtalk/knowledgebase.html

or from our Web-based Technical Support forum at

http://www.mackichan.com/techtalk/UserForums.htm

You can also contact our Technical Support staff by email, telephone, or fax. We urge you to submit questions by electronic mail whenever possible in case our technical staff needs to obtain your file to diagnose and solve the problem.

When you contact us by email or fax, please provide complete information about the problem you're trying to solve. We must be able to reproduce the problem exactly from your instructions. When you contact us by telephone, you should be sitting at your computer with the program running.

Please provide the following information any time you contact Technical Support:

- The MacKichan Software product you have installed.
- The version and build number of your installation (see **Help / About...**).
- The serial number of your installation (see **Help / System Features...**).
- The version of the Windows system you're using.
- The type of hardware you're using, including printer and network hardware.
- A description of what happened and what you were doing when the problem occurred.
- The exact wording of any messages that appeared on your computer screen.

▶ **To contact Technical Support**

- Contact Technical Support by email, fax, or telephone between 8 A.M. and 5 P.M. Pacific Time:

  Internet electronic mail address: support@mackichan.com
  Fax number: 360-394-6039
  Telephone number: 360-394-6033
  Toll-free telephone: 877-SCI-WORD (877-724-9673)

Additional Information

You can learn more about *SWP* and *SW* on our website, which we update regularly to provide the latest technical information about the program. The site also houses links to other TeX and LaTeX resources. We maintain an unmoderated discussion forum and an unmoderated electronic mail list so our users can share information, discuss common problems, and contribute technical tips and solutions. You can link to these valuable resources from our home page at **http://www.mackichan.com**.

1 Tailoring Typesetting to Your Needs

The typeset appearance of your *SWP* or *SW* document depends on typesetting specifications, all set initially by the shell you use to create the document. Starting a new document with a carefully chosen shell is important; it minimizes the typesetting modifications you may have to make. If the document shell adheres closely to your typesetting requirements, you may then be able to create the perfect typeset appearance for your document just by making one or two small changes from within the program. Once you've tailored your document to your typesetting requirements, you can export it as a shell so that you can use it repeatedly. Chapter 2 "Using Document Shells" explains more about choosing an appropriate shell and creating your own shells.

The information in this chapter answers the typesetting questions we receive most often from users. The questions involve page layout, front and back matter, tables and graphics, and mathematics. The table near the end of the chapter refers you to additional information about other typesetting tasks. If you have a basic knowledge of TeX and LaTeX and are familiar with *SWP* or *SW*, this chapter may be all you need to adapt an existing document shell to your requirements.

You can find more information in the online Help system, in links from the file `OptionsPackagesLaTeX.tex` in the `SWSamples` directory of your installation, and in the TeX and LaTeX references noted in the Troubleshooting section of this chapter.

Note The instructions for the basic processes to which we refer in this chapter appear in Chapter 2.

Cautions

The techniques we suggest for modifying the typeset appearance of your document involve working from within *SWP* or *SW* rather than using an ASCII editor to work directly with the TeX and LaTeX code. The techniques involve modifying the document class options (see page 54), adding and modifying LaTeX packages (see page 65), and inserting TeX commands in your document (see page 68). The techniques should work successfully for typesetting both device independent (DVI) files and, in Version 5, for Portable Document Format (PDF) files.

The suggested techniques are general. They work with many, *but not all,* document shells. The techniques are most likely to be successful with document shells based on standard LaTeX typesetting specifications. They are less likely to be successful if you have added an unusual combination of packages to your document, because they may cause LaTeX conflicts. If your document uses a Style Editor shell, you may find that using the Style Editor to make the changes you need is easier and more successful than working from within the document. The suggestions in this chapter may not work with Style Editor documents.

We are unable to predict the effect of these modifications on documents created with LaTeX typesetting specifications that you have added to the program. In other words, we can't guarantee results. Please note that we do not support documents created with typesetting specifications not provided with our program. Note too that some LaTeX packages are distributed with the program only as a convenience or for compatibility and may not necessarily work with *SWP* or *SW* or with other packages.

Some of the suggestions involve bypassing the way LaTeX naturally works, an approach that can sometimes have unpredictable results. Other suggestions involve placing LaTeX code in the document preamble and otherwise sending commands directly to LaTeX. Any time you add TeX fields or raw TeX code to your document, you run the risk of damaging it.

Important Even seemingly small coding errors can have large and unwelcome effects. Errors may prevent compilation, or they may damage or truncate your document irrevocably. *Save a copy of your document* before you attempt any of the modifications suggested here.

Most importantly, we urge you not to attempt extensive modifications of the typesetting specifications unless you're very familiar with TeX and LaTeX.

Tailoring the Page Layout

Making small changes to the page layout may yield the perfect typeset appearance for your document. Often, the most straightforward way to affect the page layout is by modifying one or more of the class options in effect for your document. The class options control many fundamental aspects of page layout and document design, such as paper size, body text point size, title and author information, page orientation, or columns. See LaTeX Class Options on page 55 for more information.

Remember Save a copy of your document before you attempt to modify it.

Changing the Margins

Your document has two sets of margins, one for producing your document without typesetting and one for producing your document with typesetting. If you don't intend to typeset your document, you can change the margins using the page setup specifications. Access this set of specifications with the **Page Setup** command on the **File** menu. Remember that the changes you make to the margins using the page setup specifications have no effect on the typeset appearance of your document.

Margins for typesetting are generally set in the document class specifications file (.cls file); see page 53. The .cls file usually contains different margin defaults for each paper size, so changing the paper size default automatically changes the margin defaults. The LaTeX article class defines margins of about $1\frac{7}{8}$ inches on all sides when $8\frac{1}{2}$x11 inch paper is specified, but those margins change subtly when a4 paper is specified. Regardless of the paper size, you can modify the margin settings for most but not all shells by adding the *geometry* package (see page 110). See Changing the Headers and Footers on page 5 for additional information.

Fitting More Text on a Page

Without any other modifications, adding the *geometry* package to a document enlarges the margins. This happens because LaTeX gives precedence to the *geometry* package margin settings, which usually differ from those of the document class. When you add *geometry* to your standard LaTeX article class document, the package specifies a $1\frac{5}{16}$-inch margin on the right, left, and top and a 2-inch margin on the bottom.

Changing the Margin Settings

You can use the default margins for the package or specify the margins you want by adding a command to the preamble of your document.

▶ **To change the margins of a document**

1. Add the *geometry* package to your document.

2. From the **Typeset** menu, choose **Preamble**.

3. Click the mouse in the entry area.

4. On a new line at the end of the entries, type:

   \geometry{left=$w$in,right=$x$in,top=$y$in,bottom=$z$in}

   where w and x are the left and right margins, y is the top margin, and z is the bottom margin. You can use any of the usual TeX measurement units in the command:

Unit	Value
sp	Scaled point (65536 sp = 1 pt)
pt	Point ($\frac{1}{72.27}$ in)
bp	Big point ($\frac{1}{72}$ in)
dd	Didít point (0.376 mm)
mm	Millimeter
pc	Pica (12 pt)
cc	Cicero (12 dd)
cm	Centimeter
in	Inch

5. Choose **OK**.

Changing the Line Spacing

SWP and *SW* support multiple-line spacing in typeset documents. The typesetting specifications for each document shell initially determine the line spacing of documents created with the shell. Some specifications have options that change the line spacing, but others don't. Although line-spacing specifications are sometimes presented as *draft* or *manuscript* (double spacing) or as *final* or *camera-ready* (single spacing), print quality options don't necessarily imply line-spacing changes. See LaTeX Class Options on page 56 for more information about examining and changing class options.

The LaTeX *setspace* package supports changes to line spacing. With the package, you can select single, one-and-one-half, or double spacing for the document as a whole or for parts of the document. If you find a shell that otherwise meets your typesetting requirements, add the *setspace* package and then change the spacing as needed.

Some shells and some documents created with older versions of SWP and SW may require the *doublespace* package, now superseded by *setspace*. The *doublespace* package is provided with the program for compatibility. We recommend you use the *setspace* package instead.

Note that if your document was created with a Style Editor shell, these packages probably won't work. Use the Style Editor to make any necessary line spacing changes to the typesetting specifications.

▶ **To change the line spacing of the entire document**

1. Add the *setspace* package to your document.

2. If you're using Version 4.0 or later,

   a. On the Typeset toolbar, click the Options and Packages button or, from the **Typeset** menu, choose **Options and Packages**.
   b. Choose the **Package Options** tab.
   c. From the **Packages in Use** box, select **setspace** and choose **Modify**.
   d. In the **Category** box, select **Line Spacing**.
   e. In the **Options** box, select the spacing you want.
   f. Choose **OK** twice to return to your document.

   or

   If you're using an earlier version of the program,

   a. From the **Typeset** menu, choose **Preamble**.
   b. Click the mouse in the entry area.
   c. On a new line at the end of the entries, type the command that corresponds to the spacing you want: **\singlespacing**, **\onehalfspacing**, or **\doublespacing**.
   d. Choose **OK**.

▶ **To change the line spacing for a portion of a document**

1. Add the *setspace* package to your document.

2. Place the insertion point at the start of the first paragraph whose spacing you want to change.

3. Enter a TeX field.

4. In the entry area,

   - Type **\singlespacing**, **\onehalfspacing**, or **\doublespacing**, depending on the spacing you want.

or

- Type **\setstretch{x}** where *x* is a number indicating the spacing you want. For example, the command \setstretch{3} produces triple spacing.

5. Choose OK.

6. Place the insertion point where you want to return to the original spacing.

7. Repeat steps 3–5.

Changing the Font Size of Body Text

Most document classes have a default setting for the font size used for body text. You can change the setting by modifying the document class options. Note that the body text font size is used as the basis for determining many other typesetting specifications, such as script size. Note also that some typesetting specifications override the class options.

▶ **To change the body text font size with the class options**

1. On the Typeset toolbar, click the Options and Packages button or, from the Typeset menu, choose Options and Packages.

2. Choose the Class Options tab and choose Modify.

3. In the Category box, select Body text point size.

4. In the Options box, select the font size you want.

5. Choose OK.

6. Choose OK to return to your document.

Changing the Headers and Footers

The typesetting specifications automatically generate page headers and footers using information from various LaTeX counters to generate header and footer content. You may need to modify the content or format of the headers and footers or even suppress them on certain pages.

Two packages—*fancyhdr* and *geometry* (see pages 105 and 110)—can simplify many of these changes. TeX commands accomplish the rest.

Fitting Headings into Headers

If you find that a heading is too long to fit in the header, you can define a short heading to use in its place. The short title will appear in the page header and also in the table of contents of your document. This technique works for standard LaTeX typesetting specifications, but may not work for other typesetting specifications.

▶ **To define a short heading for a section**

1. Place the insertion point at the beginning of the section heading.

2. Type the short heading enclosed in square brackets.

   The heading might then look something like this:

   [New Shorter Heading]A Much Longer Heading to Announce This Section of My Document

Suppressing Headers and Footers

You can suppress the header and footer on individual pages of your document, or you can take a more extreme approach and entirely eliminate the space allotted for printing the headers and footers.

▶ **To suppress the header and footer on an individual page**

1. Place the insertion point on the page for which you want no header or footer.

   If the page has a section heading, place the insertion point after the heading.

2. Enter an encapsulated TeX field.

3. In the entry area, type **\thispagestyle{empty}** and choose OK.

   When you typeset preview your document, you may find that the typesetting specifications for the first page of a part or chapter have been defined differently from the rest of the document and have not been affected by your change. In this case, you must suppress the header and footer information on the individual pages.

   If you want to expand the amount of information you can get on a page, you can eliminate the headers and footers entirely and expand the text area into the space they ordinarily occupy. Use the *geometry* package with some TeX commands to accomplish this change.

▶ **To eliminate the header and footer space throughout the document**

1. From the Typeset menu, choose Preamble and click the mouse in the entry area.

2. On a new line at the end of the entries, type **\pagestyle{empty}**.

3. Choose OK.

4. On the Typeset toolbar, click the Options and Packages button or, from the Typeset menu, choose Options and Packages and then choose the Package Options tab.

5. Add the *geometry* package to your document.

6. In the Packages in Use box, select geometry and choose Modify.

7. In the Category box, select Headers/Footer Space.

8. In the Options box, select **No header space**, **No footer space**, or **No header or footer space**.

9. Choose **OK** twice to return to your document.

Changing the Format of Headers and Footers

Adding a line, or *rule,* under the header requires the addition of the *fancyhdr* package and a few TeX commands.

▶ **To add a rule under the header**

1. Add the *fancyhdr* package to your document.

2. Enter an encapsulated TeX field on the first page of the body of the document.

3. In the entry area, type these two commands:

   \pagestyle{fancy}

   \renewcommand{\headrulewidth}{*x*pt}

   where *x* is the point size of the rule you want.

   For reference, this is a 1-point rule ———————— and this is a 5-point rule ▬▬▬▬▬▬▬▬. If you want to remove a rule under a header, set the linewidth to zero.

4. Choose **OK**.

You may find that the space allotted for headers and footers is inadequate, but you can increase the space easily.

▶ **To increase the space available for headers and footers**

1. From the **Typeset** menu, choose **Preamble**.

2. Click the mouse in the entry area.

3. Change the header space:

   a. Add a new line at the end of the preamble entries.
   b. Type **\setlength{\headheight}{*x*}** where *x* is the height of the header you want.
   You can use any TeX measurement unit.

4. On a new line, type **\setlength{\textheight}{*x*}** where *x* is the desired text height.

   The text height should reflect the original text height set for the document less the amount you added to the header.

5. On a new line, type **\setlength{\footskip}{*x*}** where *x* is the distance from the bottom of the text to the bottom of the footer.

6. Choose **OK**.

   If you want to expand the amount of information you can get on a page, you can eliminate the headers and footers entirely and expand the text area into the space they usually occupy, as described on page 6.

Specifying Header and Footer Information

LaTeX automatically creates headers and footers from the information in the typesetting specifications and the section headings in your document. You can override the automatic headers and footers by adding the *fancyhdr* package and making some changes to the preamble or to the body of your document. Occasionally, the typesetting specifications override the page style for certain pages, especially *exceptional pages* such as the title page or the first page of a chapter or section. Instead of specifying the headers and footers for the entire document, you can specify them for a particular page by placing TeX commands in the body of the document instead of the preamble.

The instructions below explain how to specify a header that has the title of the work on the right and a footer that has the author's name on the left and the page number on the right. You can modify the commands in step 4 to create the header and footer you want.

▶ **To specify header and footer information for the entire document**

1. Add the *fancyhdr* package to your document.

2. From the **Typeset** menu, choose **Preamble**.

3. Click the mouse in the entry area.

4. At the end of the entry area, add new lines to specify the content of the right, center, and middle areas of the header and footer (any information that follows the % is a comment and is not necessary):

   \pagestyle{fancy}

   \lhead{} %Leave the left of the header empty

   \chead{} %Leave the center of the header empty

   \rhead{Title of This Document} %Display this text on the right of the header

   \lfoot{By Author} %Display this text on the left of the footer

   \cfoot{} %Leave the center of the footer empty

   \rfoot{Page:\ \thepage} %Print the page number in the right footer

   \renewcommand{\headrulewidth}{0pt} %Do not print a rule below the header

   \renewcommand{\footrulewidth}{0pt} %Do not print a rule above the footer

5. Choose **OK**.

▶ **To specify header and footer information for selected pages**

1. Add the *fancyhdr* package to your document.

2. Place the insertion point on the page for which you want to specify header or footer information.

3. Enter an encapsulated TeX field containing the lines that follow to specify the content of the right, center, and middle areas of the header and footer.

Modify the commands as necessary (any information that follows the % is a comment and is not necessary):

\thispagestyle{fancy}

\lhead{} %Leave the left of the header empty

\chead{} %Leave the center of the header empty

\rhead{Title of This Document} %Display this text on the right of the header

\lfoot{By Author} %Display this text on the left of the footer

\cfoot{} %Leave the center of the footer empty

\rfoot{Page:\ \thepage} %Print the page number in the right footer

\renewcommand{\headrulewidth}{0pt} %Do not print a rule below the header

\renewcommand{\footrulewidth}{0pt} %Do not print a rule above the footer

4. Choose OK.

Changing the Page Numbering

Page numbering is produced automatically by the typesetting specifications for your document and in particular by the document class for the shell. However, you may require subtle changes in the page numbering scheme, perhaps to reset the numbering at some point in your document, to use lowercase roman numerals instead of arabic numbers in the front matter, to move the page number elsewhere on the page, or to remove it altogether. You can change the page numbering in your document by using packages and, in some cases, by inserting TeX commands in the body of your document.

You may find additional information is helpful when you try to modify the page numbering. See the TeX and LaTeX resources noted on page 48 and the information available about the *fancyhdr* package on page 105.

Resetting the Page Number

Page numbering sometimes requires subtle changes. You may need to set an arbitrary page number at some point in your document.

▶ **To set an arbitrary page number**

1. Place the insertion point where you want the page number to be reset.

2. Enter an encapsulated TeX field.

3. In the entry area, type **\setcounter{page}{*x*}** where x is the number from which you want page numbering to begin at this point.

4. Choose OK.

Changing the Page Numbering Style

You may need to use a different style for the page numbers in one part of your document. With a LaTeX command, you can change the page numbering style to upper- or lowercase roman numerals, upper- or lowercase letters, or arabic numbers. When you change the page numbering style, LaTeX resets the page number to 1, so you may want to reset the page number after you change the style.

▶ **To change the style of the page numbering**

1. Place the insertion point on the page whose page numbering style you want to change.

2. Enter an encapsulated TeX field.

3. In the entry area, type the command corresponding to the page numbering style you want:

Command	Page numbering style
\pagenumbering{roman}	lowercase roman numerals
\pagenumbering{Roman}	uppercase roman numerals
\pagenumbering{arabic}	arabic numbers
\pagenumbering{alpha}	lowercase letters
\pagenumbering{Alpha}	uppercase letters

4. Choose **OK**.

Shells based on the standard book document class (see page 54) use roman numerals for the page numbers in the front matter of the document but arabic numbers for page numbers throughout the rest of the document. If the shell you've chosen has a different document class and therefore a different page numbering scheme, you can change the numbering by placing LaTeX commands in your document.

▶ **To use roman numerals in the front matter only**

1. From the **Typeset** menu, choose **Preamble**.

2. Click the mouse in the entry area.

3. Create a new line at the end of the preamble entries.

4. Type **\pagenumbering{roman}** and choose **OK**.

5. Place the insertion point on a line immediately after the first chapter heading in your document.

6. Enter an encapsulated TeX field.

7. In the entry area, type **\pagenumbering{arabic}** and choose **OK**.

   Remember that LaTeX will reset the page number to 1.

Moving the Page Number

If your document style places the page number in a less than ideal spot, you can move it with the *fancyhdr* package. The package defines the content of the right, center, and left sections of the header and footer. You can place the page number in one of these areas. For more information about the *fancyhdr* package, see the online Help system and the information on pages 5 and 105.

Because your document shell may already define the content of some header and footer areas, you may need to redefine them before your changes will work correctly. For example, if you want to remove the page number from the bottom center of the page, you must redefine the area as blank so the number doesn't appear there. The instructions below explain how to move a page number from the center bottom of the page to the top right. Use them as a guideline for moving page numbers to any part of the header or footer.

▶ **To move the page number from the bottom center to the top right of the page**

1. Add the *fancyhdr* package to your document.

2. Move the page number:

    a. In the body of the document, enter an encapsulated TeX field.
    b. In the entry area, type **\pagestyle{fancy}** to establish a new page style.
    c. On the next line, add the command **\fancyhf{}** to clear the header and footer.
    d. On the next line, add the command **\rhead{\thepage}** to force the page number to the top right of the page.
    e. If you want to remove the automatically created rule under the header, add the command **\renewcommand{\headrulewidth}{0pt}** on the next line.
    f. If you want to remove the automatically created rule above the footer, add the command **\renewcommand{\footrulewidth}{0pt}** on the next line.
    g. Choose OK.

    or

    a. From the Typeset menu, choose Preamble, and click the mouse in the entry area.
    b. Start a new line at the end of the preamble entries.
    c. Type the commands specified in steps b–f, above, then choose OK.

Removing the Page Number

If the typesetting instructions for your document specify that a page number occurs on a page where you prefer not to have a number, you can suppress page numbering on the page by inserting a LaTeX command in the body of your document. Page numbering will continue as before on the next page of the document.

▶ **To suppress the page number on a page**

1. Place the insertion point on the page for which you want no page number.

2. Enter an encapsulated TeX field.

3. In the entry area, type **\thispagestyle{empty}** and choose OK.

Creating Multiple Columns

Most *SWP* and *SW* documents can be typeset in multiple columns. The typesetting specifications for each document shell determine the number of columns initially created with the shell, but you can modify the setting. Although most shells default to a single column, some default to double columns; skim through *A Gallery of Document Shells* to identify the double-column shells.

Creating double-column output for your document may require little more than changing the document class options. If you specify the double-column option, LaTeX typesets the entire body of the document in two columns. The option doesn't necessarily apply to all parts of the document front matter.

If you want finer control over which document elements are and are not typeset in multiple columns, you can use the *multicol* package to create as many as 10 columns of text, and to combine single and multiple columns on the same page. For more information, see page 121.

▶ **To create double columns**

1. On the Typeset toolbar, click the Options and Packages button or, from the Typeset menu, choose **Options and Packages**.

2. Choose the **Class Options** tab and then choose **Modify**.

3. In the **Category** box, select **Columns**.

4. In the **Options** box, select **Two columns** and choose **OK**.

5. Choose **OK** to return to your document.

▶ **To create multiple columns**

1. Add the *multicol* package to your document.

2. Define the start of the multicolumn environment:

    a. Place the insertion point where you want multiple columns to begin.
    b. Enter an encapsulated TeX field.
    c. In the entry area, type \begin{multicols}{x} where x is the number of columns you want.
    d. Choose OK.

3. Define the end of the multicolumn environment:

    a. Place the insertion point where you want multiple columns to end.
    b. Enter an encapsulated TeX field.
    c. In the entry area, type \end{multicols}.
    d. Choose OK.

Changing the Page Orientation

You can typeset most documents using portrait or landscape orientation for all or part of the document. You can change the orientation of your entire document by changing the options for the document class or by adding the *geometry* package to the document and then modifying the package options. Because the document class and package options you specify can conflict, you must be careful not to use contradictory orientation settings. The printed output depends on your printer's capabilities and settings.

▶ **To change the document orientation**

- Use the class options:

  a. On the Typeset toolbar, click the Options and Packages button or, from the Typeset menu, choose Options and Packages.
  b. Choose the Class Options tab and choose Modify.
  c. In the Category box, select Orientation.
  d. In the Options box, select the orientation you want, and then choose OK.
  e. Choose OK to return to your document.

  or

- Use the *geometry* package:

  a. On the Typeset toolbar, click the Options and Packages button or, from the Typeset menu, choose Options and Packages.
  b. Choose the Package Options tab.
  c. Add the *geometry* package to your document.
  d. From the Packages in Use box, select geometry and choose Modify.
  e. In the Category box, select Orientation.
  f. In the Options box, select the orientation you want.
  g. Choose OK twice to return to your document.

If you need to create a section of your document in a different orientation, perhaps to contain a landscaped table, we suggest that you use the *portland* package. Packages that involve rotating text, such as the *lscape* package described on page 118, are not compatible with the TrueTeX Previewer. However, PDF viewers do support rotation, so you can use the *lscape* package to rotate text in typeset PDF files; see page 118.

▶ **To change the orientation of a single page**

1. Add the *portland* package to your document.

2. Place the insertion point where you want the orientation change to occur.

3. Enter a TeX field.

4. In the entry area, type **\landscape** or **\portrait**, depending on the orientation you want, and choose OK.

5. Place the insertion point where you want to return to the original orientation.

6. Enter a TeX field.

7. In the entry area, type \portrait or \landscape, whichever is appropriate.

8. Choose OK.

Remember that you may need to change the orientation settings for your printer so that your document prints properly. Instructions differ for each printer, but most changes originate from a Setup tab. The instructions below are guidelines only.

If you have a single page with an orientation different from that of the rest of the document, you may need to print it separately after changing the printer settings accordingly.

▶ **To change the printer settings for a different orientation**

1. If you're printing from the document window, choose Print or Print PDF from the Typeset menu.

   or

   If you're printing from a previewer, choose Print or Print Setup from the File menu.

   Note that menu commands may differ depending on your print driver and previewer.

2. Look for a Setup or Properties tab or button and choose it.

3. Select the orientation you want.

4. Choose OK to return to your document or to the previewer.

Note that you may want to change the orientation of the TrueTeX display in order to properly preview documents with landscape orientation.

▶ **To change the orientation of the TrueTeX Previewer display**

1. On the Typeset toolbar, click the Typeset DVI Preview button or, from the Typeset menu, choose Preview.

2. From the Options menu in the TrueTeX Previewer, choose Preferences.

3. Choose Page Orientation and select the orientation you want.

The TrueTeX Previewer remembers the page orientation most recently set and uses it the next time you preview a document.

Changing to a Different Paper Size

The program can typeset most documents for a variety of paper sizes. You can specify the paper size by changing the class options or by adding the *geometry* package. (The package works with most but not all shells; for more details about the *geometry* package, see page 110.) Because the document class and package options you specify can conflict, you must be careful not to use contradictory paper size settings. Printing the document successfully involves changing the settings for your printer and, of course, using the correct paper in the printer.

▶ **To change the paper size**

- Modify the class options:

  a. On the Typeset toolbar, click the Options and Packages button or, from the Typeset menu, choose Options and Packages.
  b. Choose the Class Options tab and choose Modify.
  c. In the Category box, select Paper size.
  d. In the Options box, select the paper size you want, and then choose OK.
  e. Choose OK to return to your document.

  or

- Use the *geometry* package:

  a. On the Typeset toolbar, click the Options and Packages button or, from the Typeset menu, choose Options and Packages.
  b. Choose the Package Options tab.
  c. Add the *geometry* package to your document.
  d. From the Packages in Use box, select geometry and choose Modify.
  e. In the Category box, select Paper size.
  f. In the Options box, select the paper size you want.
  g. Choose OK twice to return to your document.

▶ **To change the TrueTeX Previewer settings for paper of a different size**

1. On the Typeset toolbar, click the Typeset DVI Preview button or, from the Typeset menu, choose Preview.

2. From the Options menu in the TrueTeX DVI Previewer window, choose Preferences and then choose Page Size.

3. Select the paper size you want and choose OK.

   The TrueTeX Previewer remembers and uses the page size selected most recently.

▶ **To change the printer settings for a different paper size**

1. From the Typeset menu, choose Print or Print PDF.

2. In the Page Setup tab, scroll the list of sizes in the Paper Size box to select the size you want.

   Commands and their menu location may differ depending on your print driver.

3. Choose OK three times to close the succession of dialogs and begin printing.

Tailoring the Front and Back Matter

The typesetting specifications determine the appearance of elements in the front and back matter of your document. The title page, table of contents, appendices, and bibliography may require modification to adhere to your requirements. You can make many of these changes from within *SWP* and *SW*.

Changing the Title Page

The typesetting specifications determine which elements appear on the title page of your document. If you need different information on the title page of your book, report, or thesis, you can discard the automatically generated front matter and place the information you want in a titlepage environment and a series of TeX fields in the body of your document. You must provide all required formatting for the information on the new page. Also, because the titlepage environment resets the page number of the following page to 1, you may need to reset the page numbering scheme; see page 9.

▶ **To create a customized title page**

1. At the beginning of the body of your document, enter an encapsulated TeX field.

2. In the entry area, type \begin{titlepage}.

3. Choose **OK**.

4. Enter the information for the title page, using tags and spacing commands to format the information according to your requirements.

5. At the end of the information, enter an encapsulated TeX field.

6. In the entry area, type \end{titlepage}.

7. Choose **OK**.

8. If you want to insert a table of contents or another front matter element such as a list of figures, or list of tables, enter an encapsulated TeX field.

9. In the entry area, type the TeX command for the element.

10. Choose **OK**.

    These commands include \tableofcontents, \listoffigures, and \listoftables.

11. On the Typeset toolbar, click the Front Matter button 📖 or, from the **Typeset** menu, choose **Front Matter**.

12. Delete everything and choose **OK**.

13. Save and preview the document, setting the number of LaTeX passes to 3 if your document has a table of contents.

Changing the Table of Contents

The typesetting specifications define the way each section of your document is presented in the table of contents. They determine which section headings appear in the contents, whether unnumbered headings are included, and how certain heading are listed. Changing these specifications from within the program is straightforward.

Changing the Section Levels Displayed in the Table of Contents

The tocdepth counter setting in the typesetting specifications determines which section headings are included in the table of contents. LaTeX uses *section* as level 1, regardless of the document class. For information about which section headings carry numbers, both in the table of contents and the text, see page 25.

▶ **To modify the level of section headings reflected in the table of contents**

1. From the Typeset menu, choose Preamble and click the mouse in the entry area.

2. On a new line at the end of the entries, type **\setcounter{tocdepth}{*x*}** where *x* is the deepest heading level you want to display.

3. Choose OK.

Creating Page Breaks in the Table of Contents

LaTeX generates the entries in a typeset table of contents from the section headings in your document. You can force a table of contents entry to move from the bottom of one page to the top of the next, or vice versa. Use LaTeX commands in the body of your document to lengthen pages and create page breaks in the table of contents.

▶ **To force an entry to the next page of the table of contents**

1. Compile your document to generate a table of contents.

2. Determine the table of contents entry you want to force to the next page.

3. In the body of your document, find the corresponding section heading.

4. Place the insertion point immediately above the heading.

5. Enter a TeX field.

6. In the entry area, type **\addtocontents{toc}{\protect\pagebreak}**.

7. Choose OK.

   When you compile your document to create a new table of contents, LaTeX places a page break before the table of contents entry for the section heading, and forces the entry to the next page of the table of contents. The body of your document remains unchanged.

▶ **To force an entry to the previous page of the table of contents**

1. Compile your document to generate a table of contents.

2. Determine the table of contents entry you want to force to the previous page.

3. In the body of your document, find the corresponding section heading.

4. Place the insertion point immediately above the heading and enter a TeX field.

5. In the entry area, type **\addtocontents{toc}{\enlargethispage*{1000pt}}** where the actual point value depends on how much information you're trying to fit on the page.

6. Choose OK.

7. Place the insertion point in the paragraph following the heading and enter a TeX field.

8. In the entry area, type **\addtocontents{toc}{\protect\pagebreak}** and choose OK.

   When you compile your document to create a new table of contents, LaTeX forces the entry for the section heading to the previous page of the table of contents and places a page break after the entry. The body of your document remains unchanged.

Listing Unnumbered Sections

Generally, the specifications exclude unnumbered sections, such as the introduction or acknowledgments, from the table of contents. You can force LaTeX to include them.

▶ **To include an unnumbered section in the table of contents**

1. Place the insertion point just after the unnumbered heading.

2. Enter an encapsulated TeX field.

3. In the entry area, type **\addcontentsline{toc}{*sectionlevel*}{*Name*}** where *sectionlevel* is the level of the heading you want to include, such as chapter or section, and *Name* is the text you want to appear in the table of contents.

4. Choose OK.

Listing the Bibliography and Index

The headings for two sections—the bibliography and the index—are both unnumbered and automatically generated. They typically appear in the body of the document but not in the table of contents. You can create a contents entry for both sections.

▶ **To add the bibliography or index to the table of contents**

1. Create the bibliography list or index entries.

2. Typeset compile the document to generate the index and, if you're using BibTeX, the bibliography.

3. Create the table of contents entry:
   - If you have Version 4.0 or later, add the *tocbibind* package to your document. The package automatically includes the bibliography and the index in the table of contents.

   or

   - If you have an earlier version,
     i Place the insertion point just before the index or the gray box for the BIBTEX bibliography.
     ii Enter an encapsulated TEX field.
     iii In the entry area, type **clearpage** if you're working in an article or report shell or **cleardoublepage** if you're working in a book shell.
     iv Press ENTER.
     v Type **addcontentsline{toc}**{*sectionlevel*}{*Name*} where *sectionlevel* is the level of the heading you want to include, such as chapter or section, and *Name* is the text you want to appear in the table of contents.
     vi Choose OK.

Listing the Appendices

Typesetting specifications can automatically format certain table of contents entries. For example, some specifications create contents entries for each appendix using the appendix number only; others precede the number with the word *Appendix*. If your document shell uses the appendix number only in the table of contents, you can modify the entry by adding the word *Appendix*.

▶ **To force the word *Appendix* to appear in the table of contents**

1. Place the insertion point where you want the appendix title to appear in the document.

2. Enter an encapsulated TEX field.

3. In the entry area,
   - If you're using an article shell, type:
     section***{Appendix *x*. *The Title of the Appendix*}
     where *x. The Title of the Appendix* indicates the number and title of the appendix.

   or

   - If you're using a book shell, type
     chapter***{*Appendix x. The Title of the Appendix*}.

   This command removes the automatic generation of the appendix number and changes the appearance of the title in the text to what you have entered inside the curly braces.

4. Type a second command to add the now unnumbered section to the table of contents:

   addcontentsline{toc}{section}{*Appendix x. The Title of the Appendix*}.

   If you're using a book or report shell, substitute the word *chapter* for the word *section*.

5. Choose OK.

Creating an Unnumbered Manual Bibliography

At times, you may want to customize the typeset appearance of items in a manual bibliography and the citations that refer to them. For example, you may want to remove the numbers from the bibliography item list, alphabetize the items themselves, or create citations with certain content, such as the author's name and year of publication.

▶ **To create an unnumbered manual bibliography**

1. Create your document using one of the standard LaTeX shells.

2. From the **Typeset** menu, choose **Bibliography Choice**, check **Manual Entry**, and choose **OK**.

3. Remove the typesetting of the labels from the bibliography item list:
   a. From the **Typeset** menu, choose **Preamble**.
   b. Click the mouse in the entry area and add these lines to the end of the preamble:
   \makeatletter
   \def\@biblabel#1{\hspace*{-\labelsep}}
   \makeatother
   c. Choose **OK**.

4. Create the bibliography item list, giving a label to each item:
   a. Place the insertion point where you want the bibliography item list to begin.
   b. Apply the Bibliography Item tag.
   c. In the **Key** box, type a key for the item.
   d. In the **Label** box, type the citation information as you want it to appear in the text. The typesetting specifications define the appearance of the label, adding brackets automatically as necessary.
   e. Choose **OK**.
   f. Type the bibliography item as you want it to appear in the list and press ENTER.
   g. Repeat steps c–f for each item in the list, and then press F2 to exit the list.

5. Manually alphabetize the bibliography items.

6. Create citations in the text:
   a. Place the insertion point where you want a citation to appear.
   b. On the Typeset Objects toolbar, click ![icon] or, from the **Insert** menu, choose **Typeset Object** and then choose **Citation**.
   c. In the **Citation** dialog, enter the key of the item you want to cite and choose **OK**.

7. Typeset your document.

   LaTeX places the label for the bibliography item in the citation but omits it from the bibliography list.

Creating an Appendix for Each Chapter

Creating a single appendix for your document requires no special procedures, but creating an appendix for each chapter of a book or for each section of an article requires the addition of the *appendix* package (see page 86). The package implements a subappendices environment to contain the appendix.

▶ **To create subappendices**

1. Add the *appendix* package to your document.

2. Begin the subappendices environment:
   a. Place the insertion point where you want the appendix to appear.
   b. Enter an encapsulated TeX field.
   c. In the entry area, type **\begin{subappendices}** and choose OK.

3. Type the title of the appendix and apply the section tag (for appendices within chapters) or the subsection tag (for appendices within sections).

4. Type the content of the appendix.

5. End the subappendices environment:
   a. Place the insertion point where you want the appendix to end.
   b. Enter an encapsulated TeX field.
   c. In the entry area, type **\end{subappendices}** and choose OK.

The subappendices you create are numbered with the chapter or section number and an uppercase alphabetic letter, as in 2.A, 2.B, and so on. You can modify the numbering scheme to remove the chapter or section number.

▶ **To modify the subappendix numbering scheme**

1. From the Typeset menu, choose Preamble and click the mouse in the entry area.

2. On a new line at the end of the preamble entries, type:

   \renewcommand{\setthesection}{\Alph{section}}

   or

   \renewcommand{\setthesubsection}{\Alph{subsection}}

3. Choose OK.

Creating Two Separate Indices

Some publications require that you include both a subject index and an author index in your document. You can generate a double index if you use both the *makeidx* and *nomencl* packages (see pages 119 and 121).

▶ To generate two indices

1. Add the *makeidx* and *nomencl* packages to your document.

2. Create standard index entries for the subject index throughout your document.

3. Create each entry for the author index:

    a. Enter an encapsulated TeX field.
    b. In the entry area, type **nomenclature**{*x*}{**pageref**} where *x* is the author name you want to appear in the list.
    c. Choose **OK**.

4. Modify the document preamble:

    a. From the **Typeset** menu, choose **Preamble** and click the mouse in the entry area.
    b. If your document meets all these conditions:
    - •. you are using the *SWP* or *SW* output filter (not the Portable LaTeX filter)
    - •. the highest division level in your document is section
    - •. the chapter division isn't used

    then

    i Save, close, and reopen the document.
    ii From the **Typeset** menu, choose **Preamble**.
    iii Click the mouse in the entry area and scroll to the bottom.
    iv After the line \input{tcilatex}, add a new line and type **let****chapter****undefined**.
    v Choose **OK**.

    c. At the end of the preamble entries, add these new lines:
    makeglossary
    makeindex
    renewcommand{**nomname**}{**Author Index**}
    d. Choose **OK**.

5. Include both indices in your document:

    a. Place the insertion point at the end of your document.
    b. Enter an encapsulated TeX field.
    c. In the entry area, type:
    printglossary
    printindex
    d. Choose **OK**.

6. Compile your document twice:

    a. Save the document.
    b. On the Typeset toolbar, click the Typeset DVI Compile button or, from the **Typeset** menu, choose **Compile**.
    c. Check **Generate an Index** and choose **OK**.

d. Run the MakeIndex program:
- With Version 4.0 or later,
    (1) From the Windows **Start** menu, choose **Run**.
    (2) In the **Open** box, enter this command on a single line (where line breaks occur in this instruction, enter a space):
        c:\swp50\TCITeX\TrueTeX\obsolete\makeindx -o
        c:\swp50\docs*filename***.gls -s**
        c:\swp50\tcitex\tex\latex\contrib\supported\nomencl\nomencl.ist
        c:\swp50\docs*filename***.glo**
        Note Change the name of the program directory as necessary.

    or
- With an earlier version of the program,
    (1) From the Windows **Start** menu, choose **Run**.
    (2) In the **Open** box, enter this command on a single line (where line breaks occur in this instruction, enter a space):
        c:\swp35\TCITeX\SWTools\bin\makeindx -o
        c:\swp35\docs*filename***.gls -s**
        c:\swp35\tcitex\tex\latex\contrib\supported\nomencl\nomencl.ist
        c:\swp35\docs*filename***.glo**
        Substitute the correct file name for the .glo and .gls files. These instructions reflect an *SWP* installation. Correct the path names, if necessary.
    (3) Choose **OK**.

7. Typeset compile the document file from outside the program.

    Note If you compile using *SWP* or *SW*, the compiler won't find the .gls file and won't include the nomenclature list in the typeset document.

    a. From the *SWP* or *SW* submenu in the Windows **Programs** list, choose the TrueTeX Formatter.
    b. Select the file and choose **OK**.
       If your document contains a table of contents or cross-references, you many need to compile it two or three times.

Tailoring the Body of the Document

The typesetting specifications determine the appearance of division headings, text, lists, footnotes and other generated elements, and other parts of the body of your document. Using LaTeX packages and TeX commands, you can modify the appearance of these various elements.

Changing the Appearance of Division Headings

The typesetting specifications for your document determine how the program treats the headings and numbers for chapters, sections, and other divisions. You may need to modify the specifications for all section headings or just for a few occurrences of a given division to achieve a different placement, numbering scheme, or naming scheme.

Repositioning Headings

The designers of most document shells carefully consider the placement of chapter, section, and other division headings. Nevertheless, you may need to change some or all of the heading placement specifications. With the commands available through the *sectsty* package (see page 131), you can change the location of a category of section headings, such as all chapter headings or all subsection headings. Note, however, that this process also removes the section number and therefore excludes the section from the table of contents. (See page 17 for information about adding the unnumbered section to the table of contents.) If your document was created with a Style Editor shell, use the Style Editor to reposition the headings.

▶ **To relocate all headings for a given section level**

1. Add the *sectsty* package to your document.

2. From the Typeset menu, choose Preamble and click the mouse in the entry area.

3. On a new line at the end of the preamble entries, type this command for each section heading level you want to move:

   sectionfont{x}

   where *sectionfont* reflects the heading you want to move (such as **allsectionsfont**, **partfont**, **chapterfont**, or **sectionfont**) and x is the command for the new heading location:

Command	Effect
\centering	Centered heading
\raggedleft	Set flush right with a ragged left margin
\raggedright	Set flush left with a ragged right margin

4. Choose OK.

If you need to relocate only one heading instead of an entire category of headings, you can replace the heading with a TeX command. The process maintains the section number and includes the section in the table of contents. We don't recommend this process for shells with numbered chapter headings preceded by the word *Chapter*.

▶ **To relocate a single section heading**

1. Place the insertion point where you want the heading to appear and enter an encapsulated TeX field.

2. In the entry area, type {*location**section*{*Name*}} where *location* is the location you want, *section* is the section level you want to move, and *Name* is the name of the heading as you want it to appear. You can define the location as **centering**, **raggedleft**, or **raggedright**, as described above.

3. Choose OK.

4. Delete the original heading.

Adding and Removing Section Numbering

Occasionally, you may need to extend numbering to deeper levels of section headings than the document shell specifies. On the other hand, you may need to remove the number from a heading. (See page 17 for instructions about adding the unnumbered section to the table of contents.)

▶ To extend section numbering to a different level

1. From the Typeset menu, choose Preamble and click the mouse in the entry area.

2. On a new line at the end of the preamble entries, type \setcounter{secnumdepth}{*x*} where *x* is the deepest heading level you want to number.

   Remember that LaTeX uses *section* as level 1, regardless of the document class.

3. Choose OK.

▶ To remove the number from an individual section heading

- If you're using Version 4.0 or later,

  a. Place the insertion point at the beginning of the heading.
  b. Choose Properties.
  c. In the Section Properties dialog, check Unnumbered.

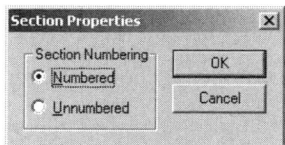

  d. Choose OK.

  or

- If you're using an earlier version,

  a. Place the insertion point where you want the heading to appear.
  b. Enter an encapsulated TeX field.
  c. In the entry area, type *section**{*Name*} where *section* is the level of the heading and *Name* is the name of the section as you want it to appear.
  d. Choose OK.
  e. Delete the original heading.

  LaTeX ignores unnumbered sections when it creates page headers. Thus, if you create an unnumbered appendix, the page headers on the appendix pages reflect the name of the previous section heading rather than the appendix.

Changing the Section Numbering Scheme

You may need to change the style of the numbers used to designate a certain section level. For example, you may want chapters to have uppercase roman numerals, sections to have capitalized alphabetic characters, and subsections to have arabic numbers. You can insert TeX commands in the preamble of your document so that your headings are numbered according to a particular scheme.

▶ **To change the numbering scheme for section headings**

1. From the **Typeset** menu, choose **Preamble** and click the mouse in the entry area.

2. At the end of the preamble entries, type \renewcommand{*level*}{*scheme*{*counter*}} where *level* is the level of the section you want to renumber, such as thechapter, thesection, or thesubsection; *scheme* is the numbering scheme, such as Roman or arabic (see page 10); and *counter* is the counter for the section level, such as part or chapter.

3. Choose **OK** to return to your document.

To produce chapters with uppercase roman numerals, sections with capitalized alphabetic characters, and subsections with arabic numbers, use these commands:
\renewcommand{\thesection}{\Alph{section}}
\renewcommand{\thechapter}{\Roman{chapter}}
\renewcommand{\thesubsection}{\arabic{subsection}}

Changing Automatic Section Names

Some typesetting specifications automatically give names, such as *Chapter* or *References,* to certain heading levels. If necessary, you can change the automatic names from, for example, *Chapter* to *Lesson* or from *References* to *Bibliography.* Making changes to the headings specifications can have wide-ranging effects, particularly if your document has a table of contents.

▶ **To change an automatically generated section name**

1. From the **Typeset** menu, choose **Preamble** and click the mouse in the entry area.

2. At the end of the preamble entries, add a new line and type

    \renewcommand{*name*}{*Newname*}

    where *name* is the title of the part of the document you want to change (see the list below) and *Newname* is the title you want.

Document part	name	Document part	name
Preface	prefacename	Table	tablename
References	refname	Part	partname
Abstract	abstractname	encl	enclname
Bibliography	bibname	cc	ccname
Chapter	chaptername	To	headtoname
Appendix	appendixname	Page	pagename
Contents	contentsname	see	seename
List of Figures	listfigurename	see also	alsoname
List of Tables	listtablename	Proof	proofname
Index	indexname	Glossary	glossaryname
Figure	figurename		

3. Choose **OK**.

The command to change the title *Contents* to *Table of Contents* is

\renewcommand{\contentsname}{Table of Contents}.

Similarly, to change *Chapter* to *Lesson,* use **\renewcommand{\chaptername}{Lesson}**; to change *References* to *Bibliography,* use **\renewcommand{\refname}{Bibliography}**. Other titles work the same way. Note that the *babel* package works successfully with redefined titles.

Each document class treats the bibliography title differently. To save work, you can use the *chbibref* package to change the title for three LaTeX document classes: article, book, and report.

▶ **To change the title of the bibliography for all document classes**

1. If you're using Version 3.51 or lower, obtain and install the *chbibref* package.

   The package is distributed with Versions 4.0 and later.

2. Add the *chbibref* package to your document.

3. From the Typeset menu, choose Preamble and click the mouse in the entry area.

4. At the end of the preamble, add a new line and type **\setbibref{*name*}** where *name* is the bibliography title you want.

   Note If you're using *babel,* place the command after the \begin{document} command by inserting it as a TeX field at the beginning of the body of the document.

5. Choose OK.

Changing the Appearance of Paragraphs

In documents based on standard LaTeX typesetting specifications, the paragraphs that immediately follow section headings are not indented, but subsequent paragraphs are indented, as in this manual.

You can remove indention for all paragraphs in your document by modifying the document preamble. However, because the beginning of unindented paragraphs is somewhat harder to see, you may also want to increase the paragraph spacing.

Removing Indention from Paragraphs

LaTeX determines the paragraph indention by the length \parindent. You can set this length to zero in the document preamble.

▶ **To remove indention for all paragraphs in the document**

1. From the Typeset menu, choose Preamble and click the mouse in the entry area.

2. At the end of the preamble, add a new line and type **\setlength{\parindent}{0in}.**

3. Choose OK.

4. Typeset preview the document to make sure the indention is correct.

Changing Interparagraph Spacing

LaTeX defines the space between paragraphs by the length \parskip. The default space is 0pt plus 1pt, which means the spacing between lines can increase by as much as 1pt, depending on how LaTeX fits the text vertically on the page. The increase translates to about $\frac{1}{10}$ to $\frac{1}{12}$ of the height of a line of text, depending on the base font point size.

If you have removed paragraph indention, a reasonable space between paragraphs is about the height of a line of text, with additional stretch or shrinkage to allow LaTeX to fit text on a page the best way possible. Therefore, if the base font size is 10pt, you might use a \parskip value of 10pt plus 1pt minus 1pt.

▶ **To modify interparagraph spacing**

1. From the Typeset menu, choose Preamble.

2. Click the mouse in the entry area and place the insertion point on a new line at the end of the entries.

3. Type **\setlength{\parskip}{*10*pt plus *1*pt minus *1*pt}**, adjusting the values according to your needs.

4. Choose OK.

5. Typeset preview the document to make sure the spacing is correct.

Changing the Appearance of Numbered Lists

The program can accept a wide variety of numbering schemes for all four levels of list items, which are designated in TeX as *theenumi* through *theenumiv*. You can change the default scheme for your entire document by entering TeX commands in the preamble or change the scheme for a portion of the document by entering commands in the body of the document. You can also use the *points* or *newpnts* package to begin the list at a number other than one.

▶ **To change the list numbering scheme for the entire document**

1. From the Typeset menu, choose Preamble and click the mouse in the entry area.

2. On a new line at the end of the entries, type this command for each list level you want to renumber: **\renewcommand{*level*}{*style*{*level_counter*}}**

    where *level* is the numbered list level you want to renumber (theenumi, theenumii, theenumiii, or theenumiv), *style* is the numbering style you want (Roman, roman, Alph, alph, or arabic), and *level_counter* is the counter for the list level (enumi, enumii, enumiii, or enumiv).

3. Choose OK.

For example, many people want list levels numbered as a standard outline: I, A, 1, a. You can achieve an outline numbering scheme with these commands:

\renewcommand{\theenumi}{\Roman{enumi}}

\renewcommand{\theenumii}{\Alph{enumii}}

\renewcommand{\theenumiii}{\arabic{enumiii}}

\renewcommand{\theenumiv}{\alph{enumiv}}

▶ **To change the list numbering scheme for a portion of the document**

1. Begin the new numbering scheme:

   a. Place the insertion point where you want the numbering scheme to begin.
   b. Enter an encapsulated TeX field.
   c. In the entry area, type this command for each list level you want to renumber:
   \renewcommand{*level*}{*style*{*level_counter*}}
   where *level* is the numbered list level you want to renumber (theenumi, theenumii, theenumiii, or theenumiv), *style* is the numbering style you want (Roman, roman, Alph, alph, or arabic), and *level_counter* is the counter for the list level (enumi, enumii, enumiii, or enumiv).
   d. Choose OK.

2. Revert to the original numbering scheme:

   a. Place the insertion point where you want the numbering scheme to revert to the default.
   b. Enter an encapsulated TeX field.
   c. In the entry area, type this command for each list level you renumbered:
   \renewcommand{*level*}{*style*{*level_counter*}}
   and designate the original style of the counter.
   d. Choose OK.

▶ **To create a numbered list that doesn't begin with 1**

1. Add the *points* or *newpnts* package to your document.

2. Place the insertion point where you want the numbered list to begin.

3. Enter an encapsulated TeX field.

4. In the entry area, type \setcounter{enumi}{*x*}\RESUME

   where *x* is a value one less than the starting number you want.

   In other words, if you want your list to begin at 10, use the command

   \setcounter{enumi}{9}\RESUME. Note that the command is case-sensitive.

5. Choose OK.

You can also use the *points* or *newpnts* package to create a numbered list interrupted by an unnumbered paragraph.

▶ To create an interrupted numbered list

1. Add the *points* or *newpnts* package to your document.

2. Place the insertion point where you want the numbered list to begin.

3. Enter an encapsulated TeX field.

4. In the entry area, type \setcounter{enumi}{*x*}\RESUME where *x* is a value one less than the starting number you want.

   The command is case sensitive. If you want your list to begin at 1, use the command \setcounter{enumi}{0}\RESUME.

5. Choose OK.

6. Begin entering the numbered list.

7. Enter the unnumbered paragraph that interrupts the list.

8. At the end of the unnumbered paragraph, enter an encapsulated TeX field.

9. In the entry area, type \RESUME and choose OK.

10. Complete the list.

Underlining and Striking Through Content

Although the program has no underlining command in the usual sense, you can add emphasis to your content with simple underlines if you use the *ulem* package. The package temporarily changes the behavior of the \em and \emph commands to support single and double underlining, wavy underlining, a single line drawn through text, and text marked over with slashes.

▶ To add simple underlines

1. Add the *ulem* package to your document.

2. Select the information you want to underline.

3. Apply the Emphasize tag to the selection.

▶ To add varied underlines and strikethroughs

1. Add the *ulem* package to your document.

2. Place the insertion point where you want the underline or strikethrough to begin.

3. Enter an encapsulated TeX field.

4. In the entry area, type *command*{*text*} where *command* is one of the following:

Tailoring the Body of the Document   31

Command	Effect
\uline	Single underline
\uuline	Double underline
\uwave	Wavy underline
\sout	Line through text
\xout	Text marked over with slashes

and *text* is the information you want emphasized.

5. Choose OK.

Stopping Hyphenation

By default, the typesetting specifications for most shells specify hyphenation. With the *hyphenat* package, you can suppress all hyphenation.

▶ **To suppress hyphenation in a document**

1. Add the *hyphenat* package to your document.

2. On the Typeset toolbar, click the Options and Packages button [icon] or, from the Typeset menu, choose Options and Packages.

3. Choose the Package Options tab.

4. In the Packages in Use box, select hyphenat and then choose Modify.

5. In the Category box, choose Hyphenation.

6. In the Options box, select None.

7. Choose OK twice to return to your document.

Note that because the *None* option prevents all hyphenation, you may get LaTeX messages about bad line breaks and overfull boxes when you typeset compile your document.

Changing Footnotes to Endnotes

Although footnotes are usually the default, the program can typeset both footnotes and endnotes for most documents. You can gather footnotes in a list at the end of your document by adding the *endnotes* package (see page 102). Instead of replacing the list of footnotes generated by LaTeX, the package stores the endnote list in a separate file with the extension .ent and deletes the list after you typeset your document.

▶ **To typeset existing footnotes as endnotes**

1. Add the *endnotes* package to your document.

2. From the Typeset menu, choose Preamble and click the mouse in the entry area.

3. At the end of the preamble entries, add a new line and type
   \renewcommand{\footnote}{\endnote}

4. Choose OK.

5. Place the endnotes list in your document:

   a. Place the insertion point at the end of your document.
   b. Enter an encapsulated TeX field.
   c. If you want the endnotes to begin on a new page, type \newpage and press ENTER.
   d. Type \begingroup and press ENTER.
   e. If you want the endnotes to be set in a normal size font instead of a smaller font, type \renewcommand{\enotesize}{\normalsize} and press ENTER.
   f. If you want to list the endnotes in the table of contents, type
      \addcontentsline{toc}{section}{Notes}
      and press ENTER.
   g. Type \theendnotes and press ENTER.
   h. Type \endgroup and choose OK.

Using Color in Documents

Although TeX was not designed to support color, you can use color in your typeset documents if your output drivers and your printer support color documents. With the *color* package (see page 97), you can specify the background color of an environment, a page, or a box. Using TeX commands, you specify which of four common color models you want to use: rgb (red, green, blue), cmyk (cyan, magenta, yellow, black), gray, or named (names known to the selected driver). The monochrome option turns off all colors and is useful if you want to preview your document using a previewer that cannot produce color.

▶ **To create a colored background for an environment**

1. Add the *color* package to your document.

2. Place the insertion point at the beginning of the environment whose background you want to produce in color.

3. Enter an encapsulated TeX field.

4. In the entry area, type \color[*model*]{*specification*} where *model* is rgb, cymk, gray, or named, and *specification* is the specific color you want, defined according to the specified model.

5. Choose OK.

For flexibility, we recommend that you leave the package options unmodified. The program sets the driver defaults using a .cfg file. If you leave the configuration unchanged, you can compile your document in another LaTeX environment without any changes.

Tailoring Theorem Environments

Together with LaTeX, *SWP* and *SW* provide for detailed formatting of mathematics. The document shells provided with the program have been designed to produce typeset mathematics that meet the theorem formatting requirements of many different publishers. However, you can use TeX commands and the *theorem* package to change the formatting slightly from within *SWP* or *SW*.

Changing Theorem Numbering

Theorems and theorem-like statements may be numbered consecutively throughout your document (Theorem 1, Corollary 2, Lemma 3, Theorem 4...) or they may be numbered independently (Theorem 1, Corollary 1, Lemma 1, Theorem 2...). Usually, the numbering scheme is established with \newtheorem statements in the document preamble. If so, you can modify the preamble to change the numbering scheme as necessary. However, sometimes the \newtheorem statements are included in the typesetting specifications for the shell. In this case, don't attempt to change the specifications. If you need a different numbering scheme for theorem environments, choose a different shell for your document.

The instructions that follow pertain only to those documents for which theorem numbering schemes are defined in the document preamble.

Creating Independent Numbering for Theorem Environments

If you have a single numbering sequence for all types of theorem environments (theorem, corollary, lemma, proposition, and so on) but you need to number them independently, you can modify the sequence with \newtheorem commands in the document preamble. The \newtheorem command has this basic syntax:

\newtheorem{*counter*}[*counter_basis*]{*counter_title*}

where *counter* is the environment to be counted (such as theorem, lemma, or corollary), *counter_basis* is the source of the count, and *counter_title* is the label for the environment. That is, the counter_basis argument determines the number of each occurrence of the environment. Thus, the command

\newtheorem{conjecture}[theorem]{Conjecture}

causes each new conjecture in your document to be labeled *Conjecture* and to be numbered with the next available number in the theorem number sequence. The command might yield a numbering sequence like this: Theorem 1, Theorem 2, Conjecture 3, Theorem 4, Conjecture 5. Removing the counter_basis argument

\newtheorem{conjecture}{Conjecture}

causes the conjectures to be numbered independently: Theorem 1, Theorem 2, Conjecture 1, Theorem 3, Conjecture 2.

In documents for which theorem-like environments are defined, the environments are usually numbered on the basis of the theorem environment. Therefore, in the preamble of the document, you might see a series of statements similar to these:

\newtheorem{theorem}{Theorem}
\newtheorem{algorithm}[theorem]{Algorithm}
\newtheorem{axiom}[theorem]{Axiom}
\newtheorem{condition}[theorem]{Condition}
\newtheorem{conjecture}[theorem]{Conjecture}
\newtheorem{corollary}[theorem]{Corollary}

⋮

To have each theorem-like environment numbered independently, you must remove the counter_basis argument—[theorem]—from each line, so that the preamble contains these lines:

\newtheorem{theorem}{Theorem}
\newtheorem{algorithm}{Algorithm}
\newtheorem{axiom}{Axiom}
\newtheorem{condition}{Condition}
\newtheorem{conjecture}{Conjecture}
\newtheorem{corollary}{Corollary}

⋮

By eliminating or varying the counter_basis argument, you can create a large variety of numbering schemes for theorem environments.

▶ To number a theorem environment independently

1. From the **Typeset** menu, choose **Preamble** and click the mouse in the entry area.

2. Scroll through the commands to find the \newtheorem statement for the environment whose numbering you want to change.

3. Remove the [counter_basis] argument from the command or replace the existing counter_basis with one that you prefer.

4. Choose OK.

Resetting the Counter for Theorem Environments

Instead of numbering theorem-like statements consecutively throughout your document, you may want to reset the numbering to 1 for each new document division (part, chapter, section, and so on). You can achieve this effect with an alternate form of the \newtheorem command that bases the numbering on a division counter instead of a theorem environment counter. The command syntax is

$$\text{\textbackslash newtheorem}\{\textit{counter}\}\{\textit{counter\_title}\}[\textit{numbered\_within}]$$

where *counter* is the environment to be counted (such as theorem, corollary, or lemma), *counter_title* is the label for the environment, and *numbered_within* is the name of an already defined counter, usually some sectional unit, such as part, chapter, or section. The *numbered_within* argument determines when the counter is reset to 1 for this particular theorem environment. Thus, the command

\newtheorem{conjecture}{Conjecture}[chapter]

causes each new conjecture in your document to be labeled *Conjecture* and to be numbered in sequence by chapter. You might see these numbers: Conjecture 1.1, Conjecture 1.2, Conjecture 2.1, Conjecture 2.2, Conjecture 2.3, Conjecture 3.1. The first digit is the chapter number and the second digit is the number of the conjecture inside the chapter.

In documents for which theorem-like environments are defined, the environments are usually numbered on the basis of the theorem environment. Therefore, in the preamble of the document, you might see a series of statements similar to these:

\newtheorem{theorem}{Theorem}
\newtheorem{algorithm}[theorem]{Algorithm}
\newtheorem{axiom}[theorem]{Axiom}
\newtheorem{condition}[theorem]{Condition}
\newtheorem{conjecture}[theorem]{Conjecture}
\newtheorem{corollary}[theorem]{Corollary}
$\vdots$

To have each theorem-like environment numbered according to a document division, you must remove the counter_basis argument—[theorem]—from each line and add the section counter indicating when the numbering should be reset. Thus, if you want to reset the numbering for theorem-like environments at the beginning of every chapter, your preamble might contain these lines:

\newtheorem{theorem}{Theorem}[chapter]
\newtheorem{algorithm}{Algorithm}[chapter]
\newtheorem{axiom}{Axiom}[chapter]
\newtheorem{condition}{Condition}[chapter]
\newtheorem{conjecture}{Conjecture}[chapter]
\newtheorem{corollary}{Corollary}[chapter]
$\vdots$

▶ **To reset the counter for a theorem environment**

1. From the **Typeset** menu, choose **Preamble** and click the mouse in the entry area.

2. Scroll through the commands to find the \newtheorem statement for the environment whose numbering you want to reset.

3. Remove the [counter_basis] argument from the command.

4. Add the [numbered_within] argument to the end of the command.

5. Choose **OK**.

Changing Theorem Formatting

Many typesetting specifications set the content of theorem environments in italics, but you can override the specifications and use upright fonts for theorems instead.

▶ **To set theorems in upright fonts**

1. Add the *theorem* package to your document.

2. From the Typeset menu, choose Preamble and click the mouse in the entry area.

3. Place the insertion point on a new line at the beginning of the preamble, before any \newtheorem statements.

4. If you want to change the font that LaTeX uses for the header of the theorem environment, type **\theoremheaderfont{*font*}** where *font* is the font family you want LaTeX to use.

5. If you want to change the font that LaTeX uses for the body of the theorem environment, type **\theorembodyfont{*font*}** where *font* is the font family you want LaTeX to use.

   To use upright fonts in the body of the theorem, type **\theorembodyfont{\upshape}**.

6. Choose OK.

Note These instructions pertain only to those documents for which theorem numbering schemes are defined in the document preamble.

Tailoring Graphics and Tables

The layout of graphics and tables depends on the typesetting specifications for your document. By using LaTeX packages in your document, you may be able to vary the layout so that you can format captions, wrap text around graphics and tables, resolve problems with certain kinds of graphics, or create landscaped tables. Graphics and tables that float sometimes require special attention, particularly if they seem to prevent LaTeX from compiling your document. TeX commands in the body of your document can force LaTeX to process all floating objects so the document can compile successfully.

Formatting Graphics and Table Captions

If your document contains graphics and tables, you may want special formatting for the caption or title. If you let the graphics and tables float, you can use the *caption2* package (see page 92) to center a multiline caption or title. (Captions and titles contained on a single line are automatically centered.) With the *caption2* package you can also change the font and size of the caption text.

LaTeX defines the space above and below the captions of floating objects (both graphics and tables) with the macros \abovecaptionskip and \belowcaptionskip. Many typesetting specifications use these macros, including sebase.cls, which defines each value as 10 pt. Although the Style Editor doesn't provide a method for changing these values, you can alter the space between floating objects and their captions by adding an external macro that is associated with the typesetting specifications for your document. Alternatively, you can change the spacing for all floating objects in a document by modifying the document preamble, and you can change the spacing for an individual floating object by placing TeX commands before and after the object.

▶ To center a multiline caption for a floating object

1. Add the *caption2* package to your document.

2. On the Typeset toolbar, click the Options and Packages button ▦ or, from the Typeset menu, choose Options and Packages.

3. Choose the Package Options tab.

4. In the Packages in Use box, select caption2 and then choose Modify.

5. In the Category box, choose Alignment.

6. In the Options box, choose Centered

7. Choose OK.

8. Choose OK again to return to your document.

▶ To change the caption font attribute and size

1. Add the *caption2* package to your document.

2. On the Typeset toolbar, click the Options and Packages button ▦ or, from the Typeset menu, choose Options and Packages.

3. Choose the Package Options tab.

4. In the Packages in Use box, select caption2, and then choose Modify.

5. In the Category box, choose Caption font size.

6. In the Option box, select the size you want: script, footnote, small, normal, large, or Large.

7. In the Category box, choose Caption label attribute.

8. In the Option box, choose the attribute you want: Upright, *Italic, Slanted,* SMALL CAPS, medium, **Bold,** Roman, Sans Serif, or `Typewriter`.

9. Choose OK.

10. Choose OK again to return to your document.

▶ To change the caption spacing for all floating objects in a document

1. Open your document.

2. From the Typeset menu, choose Preamble.

3. Click the mouse in the entry area and then start a new line.

4. Enter these lines:

   \setlength{\abovecaptionskip}{0pt}

   \setlength{\belowcaptionskip}{0pt}

5. Choose OK.

   You may need to experiment with the values in the commands to achieve the spacing you want.

▶ **To change the caption spacing for an individual floating object**

1. Place the insertion point at the end of the line before the floating object and press ENTER.

2. On the Typeset Object toolbar, click [TeX icon] or, from the Insert menu, choose Typeset Object and then choose TeX Field.

3. In the entry area, type

   \setlength{\abovecaptionskip}{0pt}

   \setlength{\belowcaptionskip}{0pt}

4. Choose OK.

   Again, you may need to experiment with the values in the commands to achieve the spacing you want.

5. To restore the default values, place the insertion point on the line after the floating object and press ENTER.

6. On the Typeset Object toolbar, click [TeX icon] or, from the Insert menu, choose Typeset Object and then choose TeX Field.

7. In the entry area, type

   \setlength{\abovecaptionskip}{10pt}

   \setlength{\belowcaptionskip}{10pt}

8. Choose OK.

Wrapping Text Around Graphics

Wrapping text around graphics and tables adds interest to the appearance of a document. You can wrap text around in-line objects at the side of the page if you use the *wrapfig* package and add several TeX commands in the body of your document. The package creates an artificial floating environment, so the graphics and tables can carry captions that appear in the text and the list of graphics. Suppose you want to wrap text around a graphic with a caption, as we do here. We have

Pi

placed the graphic on the inside edge of the page, with no overhang into the margin. We've wrapped four lines of text around the graphic, allowing room for both the graphic and the caption. To create this effect, we followed the instructions below, using this command in an encapsulated TeX field: \begin{wrapfigure}[4]{i}[0pt]{0pt}.

▶ **To wrap text around a floating object**

1. Add the *wrapfig* package to your document.

2. Place the insertion point where you want to insert the floating object.

3. Begin the wrapfigure or wraptable environment:

    a. Enter an encapsulated TeX field.
    b. In the entry area, type:
    \begin{wrapfigure}[*w*]{*x*}[*y*]{*z*}
    or
    \begin{wraptable}[*w*]{*x*}[*y*]{*z*}
    where

    w is the number of vertical lines to be narrowed to accommodate the graphic or table (optional).

    x is the placement of the graphic or table (required). Uppercase indicates *float;* lowercase indicates *exactly here*:

Position	Effect
r or R	Right side of text
l or L	Left side of text
i or I	Inside edge, near the binding (two-sided documents)
o or O	Outside edge, away from the binding (two-sided documents)

    y is the amount of overhang—the distance the graphic or table should extend into the margin (optional).

    z is the width of the graphic or table (required). If you specify a width of zero (0pt), the package uses the actual width of the graphic or table to determine the wrapping width.

    c. Choose **OK**.

4. Insert the in-line object.

5. If you want the object to have a caption or title:

    a. Enter an encapsulated TeX field.
    b. In the entry area, type **caption**{*text*} where *text* is the caption or title you want.
    c. Choose **OK**.

6. End the wrapfigure or wraptable environment:

    a. Enter an encapsulated TeX field.
    b. In the entry area, type **end{wrapfigure}** *or* **end{wraptable}**.
    c. Choose **OK**.

If your document combines both objects wrapped around text and objects that float, LaTeX may not sequence both kinds of objects correctly, although the graphic or table numbers will be correct. You may be able to correct the situation by adding the *float* package (see page 108) and revising each regular floating object—but not the wrapfigure or wraptable objects—to specify the Here placement option. See page 108 for more information about the *float* package.

Managing EPS Graphics

Documents containing Encapsulated PostScript (EPS) graphics can typeset incorrectly because the PostScript filter supplied with the program occasionally misrenders the graphics. Letters that appear in the graphics may be displaced, or the appearance of the graphic may be incorrect. You can bypass the problem using the *graphicx* package.

The *graphicx* package has options for several different typeset output drivers. When the driver option is unchanged, LaTeX typesets your document using the default driver for the current LaTeX installation. For *SWP* and *SW* installations, the default driver is tcidvi, which uses the supplied graphics filter. Thus, the graphics can be misrendered. However, if you actively choose the dvips driver option, which is the default driver for typical Unix LaTeX installations, LaTeX uses the native PostScript capabilities for the current display device. Thus, when you typeset, the graphic appears on the TrueTeX Previewer screen as a box containing the path name of the EPS file, but it appears in print correctly using the PostScript interpreter in the printer.

▶ **To bypass difficulties with EPS graphics**

1. Add the *graphicx* package to your document.

2. Save the document as a Portable LaTeX file:

   a. From the **File** menu, choose **Save As**.
   b. In the **Save as type** box, choose **Portable LaTeX (*.tex)**.
   c. Change the directory and file name as necessary.
   d. Choose **Save**.

3. On the Typeset toolbar, click ▦ or, from the **Typeset** menu, choose **Options and Packages**.

4. Choose the **Package Options** tab.

5. In the **Packages in Use** box, select **graphicx**, and then choose **Modify**.

6. In the **Options** box, select **dvips** and choose **OK**.

7. Choose **OK** to return to your document.

If your document is a Style Editor document or a LaTeX 2.09 document, it can't be saved as a Portable LaTeX file. However, you can import the contents of your document into a new document, modified as described above, and successfully bypass the EPS difficulty.

Managing Floating Objects

Documents that contain many floating objects may occasionally encounter LaTeX processing problems. When you typeset your document, LaTeX tries to process floating objects as it encounters them, anchoring them throughout the document. However, if it can't place an object because of its size or if float placement options don't fit, LaTeX will hold the floating object, and all following floating objects, until the end of the document and then generate the error "Too many unprocessed floats." You can force LaTeX to process floating objects with a TeX command in the body of your document. Alternatively, you can use the *float* package to manage the placement of floating objects.

▶ **To force the output of floating objects**

1. Place the insertion point in an appropriate location in your document, such as the end of a chapter or section.

   Because forcing the output will end the page, you may have to experiment to find the best location for the TeX command. See page 80 for information about using the *afterpage* package, which offers more flexibility in the placement of the TeX field.

2. Enter a TeX field.

3. In the entry area, type **clearpage** and choose **OK**.

   When LaTeX encounters this command as it typesets your document, it outputs any floating objects that occur in the document before the command.

   If you add the *float* package to a document and select only the *here* placement option for a floating graphic, the program automatically uses the H placement option. (This option isn't available if you are using the Portable LaTeX filter.)

▶ **To manage the output of floating objects**

1. Add the *float* package to your document.

2. Revise the properties of all floating graphics to select only the *here* option.

3. Revise the properties of all tables that float:

   Remember that a table that floats is implemented with the fragment table4_3, which has this structure:

	Head	Head	Head	
[B]	entry	entry	entry	caption [E]
	entry	entry	entry	
	entry	entry	entry	

   a. Select the TeX field named [B] and choose **Properties**.
      The program opens the field, which contains the string \\begin{table}[tbp]\\centering.
   b. Change [tbp] to **[H]**.
   c. Choose **OK**.

Creating Landscaped Tables

If your tables are too wide for portrait orientation, you can produce them on landscaped pages. Although packages that involve rotating text aren't compatible with the TrueTeX previewer provided with *SWP* and *SW*, you can use the *portland* package to create a landscaped page within a DVI file, as described on page 126. Remember that you may need to change the printer settings to print landscape pages properly. You may need to print any landscaped pages in a separate printing run. Note that you can use the *lscape* package with PDF viewers, which support rotation.

▶ **To create a landscaped table**

1. Add the *portland* package to your document.

2. Place the insertion point where you want to change the page orientation to landscape.

3. Create an encapsulated TeX field.

4. Type **landscape** and then choose OK.

5. Create the table.

6. Place the insertion point where you want to return to the original page orientation.

7. Create an encapsulated TeX field.

8. Type **portrait** and then choose OK.

Making Additional Typesetting Changes

Your typesetting tasks may differ from those discussed in this chapter. The table beginning on the next page identifies many additional tasks and suggests packages that can help you accomplish them. Most of these packages work with most *SWP* and *SW* documents; that is, the documents will compile successfully. However, you may not be able to preview the document with TrueTeX. See Chapter 3 "Using LaTeX Packages" for more information.

Making Additional Typesetting Changes   43

To	Use Package
Acronyms	
- Create a list of acronyms.	Acronym
- Ensure that all acronyms are spelled out at least once.	Acronym
Algorithms	
- Prevent algorithm statements from breaking over page boundaries.	Algorithm
- Create a list of algorithms on the table of contents page.	Algorithm
- Include complex algorithm statements in documents.	Algorithmic
AMS Documents	
- Produce documents for publication in AMS journals.	AMSMath, AMSFonts
APA Documents	
- Format citations according to APA requirements.	Apacite
- Produce documents meeting requirements of APA Publication Manual.	Apalike, Apalikeplus
Arrays	
- Format columns.	Array
- Align numbers on decimal point in tabular or array columns.	Dcolumn
Bibliographies	
- Indicate references in citations and reference lists with labels instead of numbers.	Drftcite
- Find uncited bibliography items.	Drftcite
- Print the key for each bibliography item in the margin of the page on which it occurs.	Showkeys
- Produce bibliography entries acceptable to APA journal style.	Apalike, Apalikeplus
- Improve bibliography spacing in a 2-column document.	Bibmods
- Automatically include the bibliography in the table of contents.	Tocbibind
BIBTeX Bibliographies	
- Produce BIBTeX bibliographies formatted according to *The Chicago Manual of Style, Ed. 13*.	Chicago
- Format documents for astronomy journals.	Astron
- Customize citations for seven BIBTeX bibliography styles.	Harvard
- Required with BIBTeX bibliography style newapa.bst.	Newapa
- Produce BIBTeX bibliographies with four varieties of author-date citations.	Authordate1-4
- Create separate BIBTeX bibliographies for each included file and for the document as a whole.	Chapterbib
Boxes	
- Create a boxed area on the page.	Boxedminipage
- Place boxes around content.	Fancybox
Captions	
- Rotate a figure or table caption.	Rotating
- Customize width, alignment, style, and font of captions for floating objects, including objects presented in landscape.	Caption/caption2

To	Use Package
Citations	
- Compress and order lists of numerical citations to show a range of numbers.	Cite
- Format citations according to APA requirements.	Apacite
- Indicate references in citations and reference lists with labels instead of numbers.	Drftcite
- Compress, order, and superscript sorted lists of numerical citations.	Overcite
Color	
- When the installed dvi driver supports color, produces boxes or entire pages with colored backgrounds.	Color
Columns	
- Improve bibliography spacing in a 2-column document.	Bibmods
- Format text in up to 10 columns.	Multicol
- In multi-column documents, balance the final columns of text for an attractive appearance.	Multicol
- In 2-column documents, save marks from first column.	Fix2col
Cross-references	
- Print the key for each cross-reference in the margin of the page on which it occurs.	Showkeys
- Create cross-references to labels outside the document.	Xr
Double-sided Printing	
- Specify margin offsets for two-sided printing.	Geometry
Double spacing	
- Specify double spacing.	Setspace
Endnotes	
- Create end notes instead of footnotes.	Endnotes
Equations	
- Enhance the typeset appearance of displayed equations, sub- and superscripts, and other mathematical constructs.	AMSMath
- Change placement of equation numbers and tags.	AMSMath
- Change placement of equations in displayed mathematics.	AMSMath
Exam Questions	
- Print point value of questions in margin.	Newpnts
- Print point value of questions in margin.	Points
Exercises	
- Bind a solution to its exercise.	Answers
Figures and Graphics	
- Rotate a figure.	Rotating
- Include graphics in documents.	Graphicx
- Produce small graphics in larger floating graphics or tables.	Subfigure

To	Use Package
Floating Objects	
- Place floating elements on the next page output.	Afterpage
- In 2-column documents, keep floating elements in sequence.	Fix2col
- Ensure that a floating element appears in print only after reference to it.	Flafter
- Produce small graphics and tables within larger floating tables and graphics.	Subfigure
- Define floating objects.	Float
Fonts	
- Scale a font up or down.	Scalefnt
- Substitute PostScript fonts.	PSNFSS
- Use the Times font for text but leave mathematics in CM fonts.	Times
- Insert symbol fonts for Blackboard bold, Fraktur.	AMSFonts, AMSSymb
- Change uppercase text to lowercase, or lowercase to uppercase, leaving certain LaTeX elements unchanged.	Textcase
- Sample the appearance of a font family.	Fontsmpl
Fonts for Mathematics	
- Use the AMS Euler fonts in mathematics.	Euler
- Use Times fonts for text and mathematics and to provide ligatures and improved kerning.	Mathtime
- Insert AMS symbol fonts for Blackboard bold and Fraktur.	AMSFonts
- Use large symbols.	Exscale
- Use 11 symbols not normally available: ℧ ⋈ □ ◊ ⇝ ⊏ ⊐ ◁ ◃ ▷ ▹.	Latexsym
- Use Feynman slashed character notation.	Slashed
Footnotes	
- Place footnotes in tables.	Blkarray
- Customize the appearance of footnotes.	Footmisc
- In 2-column documents, print all footnotes at foot of right-hand column.	Ftnright
- Create end notes instead of footnotes.	Endnotes
Headers and Footers	
- Customize content and format of headers and footers.	Fancyhdr
- Specify headers and footers.	Geometry
- In 2-column documents, save marks from first column.	Fix2col
Headings	
- Modify the appearance of division headings.	Sectsty
Hypertext Links	
- Make cross-references into hypertext links for PDF output.	Hyperref
Hyphenation	
- Handle hyphenation, punctuation, and other language-specific issues for non-English documents.	Babel
- Disable hyphenation in selected parts of a document.	Hyphenat

To	Use Package
Indention	
- Indent the first line of all sections.	Indentfirst
Indexes	
- Create a document index.	Makeidx
- Print index commands in the margin of the page on which they occur.	Showidx
- Automatically include the index in the table of contents.	Tocbibind
Labels	
- Print the key for each label in the margin of the page on which it occurs.	Showkeys
- Print all keys and markers in the margin of the page on which they occur.	Showlabels
LaTeX	
- Format LaTeX counters with a separator every three digits.	Comma
- Prevent counters from being reset.	Remreset
- In 2-column documents, save marks from first column.	Fix2col
Limit Placement	
- Change placement of limits for integrals, operators, summations, and other symbols.	AMSMath
Line Breaks	
- Break path statements, URLs, and email addresses for better line breaks.	Url
Line Numbers	
- Add line numbers to paragraphs.	Lineno
Line Spacing	
- Produce double, single, or one-and-a-half spacing.	Setspace
List of Figures/List of Tables	
- Automatically include the list of figures and list of tables in the table of contents.	Tocbibind
List of Symbols	
- Create a nomenclature list.	Nomencl
Lists	
- Change spacing for list items.	Paralist
Lowercase Text	
- Change uppercase text to lowercase, leaving certain LaTeX elements unchanged.	Textcase
Margins	
- Customize margins.	Geometry
Mathematics	
- Enhance the typeset appearance of displayed equations, sub- and superscripts, and other mathematical constructs.	AMSMath
Non-English Documents	
- Manage language-specific issues for non-English documents.	Babel

To	Use Package
Numbered Lists	
- Change the style of the counter for numbered lists.	Enumerate
Page Breaks	
- Break tables between pages.	Xtab
Page Layout	
- Customize page layout (margins, orientation, paper size, headers, footers, etc.).	Geometry
- Draw an illustration of the LaTeX layout of the current document.	Layout
Page Numbers	
- Remove page numbers from opening and other pages.	Nopageno
Page Orientation	
- Rotate content.	Fancybox
- Rotate parts of a page.	Graphicx
- Rotate text 90 degrees.	Lscape
- Change page orientation (portrait, landscape).	Geometry
Page References	
- Print the key for each page reference in the margin of the page on which it occurs.	Showkeys
- Automatically enhance page references with text such as "on the next page" or "on the facing page".	Varioref
- Create page references to labels outside the current document.	Xr
Paper Size	
- Change paper size.	Geometry
Punctuation	
- Manage punctuation and other language-specific issues for non-English documents.	Babel
Scaling	
- Scale parts of a page.	Graphicx
Strikethroughs	
- Strike through text with lines or slashes.	Ulem
Symbols	
- Use large symbols.	Exscale
- Define the lambdabar λ and other symbols.	Revsymb
- Display characters properly in text editors.	Inputenc
- Use 11 symbols not normally available: ℧ ⋈ □ ◇ ⇝ ⊏ ⊐ ◁ ⊴ ▷ ⊵.	Latexsym
- Use Feynman slashed character notation.	Slashed
Table of Contents	
- Automatically include front and back matter in the table of contents.	Tocbibind

To	Use Package
Tables	
- Change table headings on last page.	Xtab
- Control horizontal lines in tables.	Hhline
- Create environments similar to array and tabular.	Blkarray
- Break long tables between pages.	Longtable
- Create footnotes in tables.	Longtable
- Produce colored backgrounds and rules for table columns and rows.	Colortbl
- Align numbers on decimal point in tabular or array columns.	Dcolumn
- Create tables longer than one page.	Ltxtable
- Rotate text 90 degrees.	Lscape
- Rotate a table.	Rotating
- Produce small tables within larger floating tables or graphics.	Subfigure
- Create tables longer than a page.	Supertabular
- Create a table with a specified width.	Tabularx
- Break tables between pages.	Xtab
- Format columns in tabular environments.	Array
Theorems	
- Customize the appearance of theorem and theorem-like environments.	Theorem
Underlining	
- Underline text with single, double, or wavy lines.	Ulem
Uppercase Text	
- Change lowercase text to uppercase, leaving certain LaTeX elements unchanged.	Textcase
Verbatim Representations	
- Include programming and other verbatim statements in documents.	Alltt
- Display information exactly as it is entered at a terminal.	Verbatim
Wrapped Text	
- Wrap text around floating elements positioned at the side of the page.	Wrapfig

Troubleshooting

If your document doesn't properly compile, you must find and correct the problem. A thorough knowledge of TeX and LaTeX is invaluable in this process. In addition to the information we provide here, we encourage you to seek additional enlightenment from the resources listed on page viii. These excellent resources contain lists of error messages and careful explanations of possible causes. They also describe in detail how you can attempt to recover from errors in TeX and LaTeX as your file is being processed. In addition, you can find helpful information from the Usenet news group at comp.text.tex.

Resolving LaTeX Errors

The simplest errors to find and fix result from mistyping or omitting commands. In particular, LaTeX reacts negatively when it encounters these conditions:

- Misspelled commands or environment names, as in \beginn{wraptable}.
- Improperly matched or missing braces or delimiters, as in \begin{multicols]{3} or \begin{multicols}{3.
- Improperly using a character with a special meaning in TeX, such as #, %, &, or \.
- Missing \end commands.
- Missing command arguments.

To minimize errors of this type, take the time to proofread when you type TeX commands in TeX fields, in the preamble of your document, or in dialogs that pass your commands directly to LaTeX.

These and other errors are displayed in the LaTeX and PDFLaTeX windows as your document is being processed and are also recorded in the .log file that is created each time you typeset the document. If LaTeX or PDFLaTeX finds an error in your document, the window displays an error message, like this:

```
Runaway argument?
! Paragraph ended before \multicols was complete.
<to be read again>
                        \par
l.70

?
```

The message will help you identify and correct the problem.

The error messages contain much helpful information. The exclamation point signals the error and the information on that line describes the nature of the problem. The line number (line 70 in the illustration above) indicates approximately where in your document or in a related typesetting specifications file the error has occurred. Because lines in an *SWP* or *SW* document don't correspond to lines in the .tex file, you may need to use an ASCII editor to locate the error precisely. Often, the message includes a portion of the document text, which may help you find the error.

If all activity ceases, the compiler may be waiting for you to tell it what to do. You can try to ignore the problem, solve it and keep going, or halt the typesetting process. For example, the question mark in the last line of the message above indicates that the compiler has a question. If you type ? and press ENTER, the compiler responds with this message:

```
Type <return> to proceed, S to scroll future error messages,
R to run without stopping, Q to run quietly,
I to insert something, E to edit your file,
1 or ... or 9 to ignore the next 1 to 9 tokens of input,
H for help, X to quit.
?
```

Now you can choose a course of action. After processing stops, open the .log file to examine the messages at a more leisurely pace.

Compare the .log file to the .tex file. Pay particular attention to any errors that occur in the preamble. Incorrect definition statements or statements that hide environment changes can cause compilation errors.

▶ **To isolate a LaTeX error**

1. Make a copy of the .tex file.

2. Open the .tex file with an ASCII editor and scroll to the area in which the error occurs.

3. Place a percent sign at the beginning of each line to create comments ignored by the formatter.

4. Use the TrueTeX Formatter to compile the document:

   a. From the *SWP* or *SW* program group, choose the TrueTeX Formatter.
   b. Select the file and choose OK.

5. Repeat steps 4 and 5, commenting out larger and larger portions of the document, until LaTeX handles the document correctly.

   The area that is commented out contains the error.

6. Beginning from the top of the commented portion of the file, remove the percent signs from several lines of the file.

7. Recompile the document.

8. Repeat steps 7 and 8 until you isolate the cause of the failure.

9. Search for a mistake in that portion of the document.

Repairing a Damaged Document

If your document has been damaged or become corrupted in some way, you may be able to repair it if you can isolate the problem. However, not all documents can be repaired and read successfully. The program does not handle every possible construct that might occur in a native LaTeX document. If your document was originally written in native LaTeX and then imported to *SWP* or *SW,* some problems may persist. In general, if a LaTeX document contains a construct that differs from Plain TeX (such as `array` versus `matrix`), you may be able to open the document in *SWP* or *SW* if you modify it to use the LaTeX construct.

▶ **To repair a damaged document**

1. Make a copy of the .tex file.

2. Use the TrueTeX formatter to compile the document.

3. Isolate and correct any LaTeX errors, and then recompile the document.

4. When LaTeX compiles the document successfully, try to open the document in *SWP* or *SW*.

5. As the program loads the document, carefully watch the display of paragraph numbers on the Status bar.

6. If the document loads successfully, you have repaired the document successfully. If not, note the paragraph number where the error occurred.

7. Open the document with an ASCII editor.

8. Isolate the error, as described in Resolving LaTeX Errors, above.

Because errors that occur at this level are usually document-specific, they don't lend themselves to the kinds of general suggestions we offer in this manual. However, you may be able to make the document work in *SWP* or *SW* by enclosing the lines that cause the error in an encapsulated TeX field. If so, consider it a temporary fix and pursue a more robust solution to the problem.

Important Encapsulating erroneous code is a temporary workaround.

▶ **To encapsulate information in a TeX field**

1. Open the document with an ASCII editor.

2. If much of the document is still commented out, remove the percent signs from all but the few lines surrounding and containing the error.

3. Copy the commented lines to the clipboard.

4. Save and close the file.

5. Open the document in *SWP* or *SW*.

6. Place the insertion point where the commented lines should appear.

7. Copy the lines to an encapsulated TeX field:

   a. Enter an encapsulated TeX field.
   b. Paste the lines from the clipboard.
   c. Remove the percent sign at the beginning of each line.
   d. Choose **OK**.

8. Typeset compile the document.

9. If the document compiles correctly, you've successfully repaired the document.

    If the document still doesn't compile correctly, remove the TeX field from the document.

Complex \def and \renewcommand statements in the document preamble can prevent the program from loading the document. If your document preamble contains such commands, you may be able to read the file if you place the statements in an external file and then input the file from the preamble.

▶ **To place statements in an external file**

1. Open the document with an ASCII editor.

2. Select the statements you want to place in another file, and cut them to the clipboard.

3. In place of the deleted lines, type **input{*filename.tex*}** where *filename* is the name of the ASCII file you will create in steps 4–6.

4. Open a new file with an ASCII editor.

5. Paste the statements from the clipboard into the file.

6. Name the file with a .tex extension.

7. Save the file in an appropriate subdirectory of the TCITeX/TeX directory of your program installation.

    The file must be in a TCITeX/TeX subdirectory or LaTeX won't find it when you try to compile the document.

8. Try to open and compile the original document in *SWP* or *SW*.

2 Using Document Shells

Every *SWP* and *SW* document is created from a template called a *document shell*. Each shell carries several sets of specifications that determine its fundamental structure and appearance. Those specifications, and the structure and appearance they define, extend to each new document you create with a given shell.

The `Shells` directory of your program installation contains over 150 document shells that you can use to create books, exams, articles, reports, letters, theses, faxes, and other documents. The shells have the extension .shl.

Although many are similar, no two shells are exactly alike. Some shells create documents with a structure and components common to books; other shells create documents with a structure and components common to theses, reports, or articles. Certain shells provide for front matter that includes only a short title section; others provide a title page, table of contents, list of figures, list of tables, acknowledgments, and preface. Some shells create double-spaced, single-column pages; others create single-spaced, double-column pages. Many, but not all, shells provide item tags for theorem environments—such as theorems, lemmas, corollaries, propositions, and conjectures.

This chapter explains the role of shells and their associated specifications in the creation and typesetting of *SWP* and *SW* documents. It explains how to choose a suitable shell, how to tailor a shell to your needs, and how to create a new shell for subsequent use. The chapter also explains how to install and use typesetting specifications not provided with the program but obtained from an outside source, such as a publisher.

Understanding Document Shells

Each document shell is associated with several sets of specifications. One set, consisting of page setup specifications, print options, and a style file with a .cst extension, determines the appearance of your document when you preview or print it without typesetting or when you display it in the document window. *These specifications have no effect on the typeset appearance of your document* and they aren't our focus here.

The other set—the typesetting specifications—are our primary concern. The specifications determine the document class and fundamental structure of the shell and may also specify LaTeX *packages*—sets of additional typesetting instructions—that provide a particular typesetting capability. Typesetting specifications govern all aspects of the typeset appearance of the shell and any documents created with the shell. They govern type face; type size; margins; page size; line spacing; location and appearance of headers, footers, and section headings; paragraph layout; indention; page breaks; automatic generation of cross-references, table of contents, and other document elements; and other typographic details too numerous to mention. Most of the specifications are contained in LaTeX formatting files with extensions of .cls, .clo, and .sty, although others may be contained in your document. Regardless of whether you create a DVI file or, in Version 5, a PDF file when you typeset, this collection of specifications governs the typeset appearance of your document.

> **Note** Before Version 3.5, we used the word *style* to refer to the typesetting specifications. In newer versions of LaTeX, the *document style* has been renamed the *document class*. We now use the word *style* to refer to the .cst file and not to the typesetting specifications.

When you create a new document, you select a document shell. The program opens a new document and copies the shell into it, along with the shell's typesetting specifications, style, page setup, and print options. Until you change it in some way, the new document is identical to the shell. It has the same class and structure, uses the same LaTeX packages, and produces the same appearance in print.

LaTeX Document Classes

The document class named in the typesetting specifications determines the basic structure of the shell and of any documents you create with it. The class specifies the kind of document to be produced and defines its general structure as a book, report, article, or other kind of document. The class also determines the elements, environments, and constructs allowed in the document. Document class files have an extension of .cls.

About half of the shells provided with the program have standard LaTeX base classes; they are created with book.cls, report.cls, or article.cls. Although many of the other shells produce similar kinds of documents, they have different, more specialized base classes, as we see with the shells that produce articles formatted for a specific journal or theses formatted to meet the requirements of a particular university.

The rest of the shells have the base class sebase; they were developed with the Style Editor. Using the Style Editor itself to modify Style Editor typesetting specifications is usually more efficient than working with from within *SWP* or *SW*. The typesetting techniques we suggest in this manual don't necessarily apply to Style Editor shells or to the documents created with them.

▶ **To determine the document class for a document**

1. On the Typeset toolbar, choose the Options and Packages button ▣ or, from the **Typeset** menu, choose **Options and Packages**.

2. Choose the **Class Options** tab.

   The first line of information indicates the class; in this example, the class is article:

   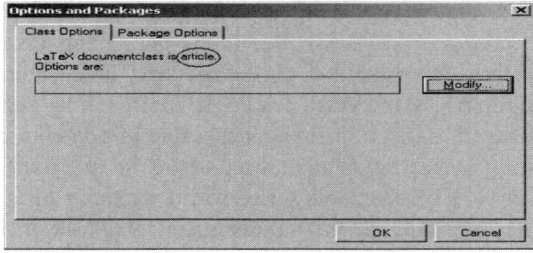

3. Choose **OK** to return to the document.

LaTeX Class Options

Although the document class defines a shell document in broad typesetting terms, LaTeX needs additional typesetting instructions to format a document completely. Some of these instructions come from *document class options,* a collection of formatting instructions that define typesetting in more detail. The options can control body text font size, page orientation, number of text columns, print quality, page size, and many other aspects of document design and typesetting.

Default Class Options

Usually, each class option has several available settings, one of which has been selected as the default, such as *10 point* for the body text font size or *portrait* for the page orientation. The options and their corresponding defaults differ from class to class. The default settings for the three LaTeX base document classes—article, book, and report—appear below. The class option defaults are in effect unless otherwise noted in the shell.

Class Option Defaults for article.cls

Category	Default	Options
Body text point size	10 pt	11 pt, 12 pt
Paper size	8.5x11	a4, a5, b5, Legal size, Executive size
Orientation	Portrait	Landscape
Print side	One side	Both sides
Quality	Final	Draft
Title page	No title page (Title area on page 1)	Title page
Columns	One	Two
Equation numbering	On right	On left
Displayed equations	Centered	Flush left
Bibliography style	Closed	Open
Babel language	English U.S.	See page 89.

Class Option Defaults for book.cls

Category	Default	Options
Body text point size	10 pt	11 pt, 12 pt
Paper size	8.5x11	a4, a5, b5, Legal size, Executive size
Orientation	Portrait	Landscape
Print side	Both sides	One side
Quality	Final	Draft
Title page	Title page	No title page
Columns	One	Two
Start chapter on left	No	Yes
Equation numbering	On right	On left
Displayed equations	Centered	Flush left
Open bibliography style	Closed	Open
Babel language	English U.S.	See page 89.

Class Option Defaults for report.cls

Category	Default	Options
Body text point size	10 pt	11 pt, 12 pt
Paper size	8.5x11	a4, a5, b5, Legal size, Executive size
Orientation	Portrait	Landscape
Print side	One side	Both sides
Quality	Final	Draft
Title page	Title page	No title page
Columns	One	Two
Start chapter on left	No	Yes
Equation numbering	On right	On left
Displayed equations	Centered	Flush left
Open bibliography style	Closed	Open
Babel language	English U.S.	See page 89.

The option settings differ from shell to shell. If you're using a shell based on a customized document class, the default categories and the corresponding options may differ. You can learn which document class options are in effect for a particular shell.

▶ **To examine the class option defaults**

1. Open a document created with the shell.

2. On the Typeset toolbar, choose the Options and Packages button ▦ or, from the **Typeset** menu, choose **Options and Packages**.

3. Choose the **Class Options** tab.

   If settings other than the document class defaults are in place, the program displays them in the **Options** box, like this:

   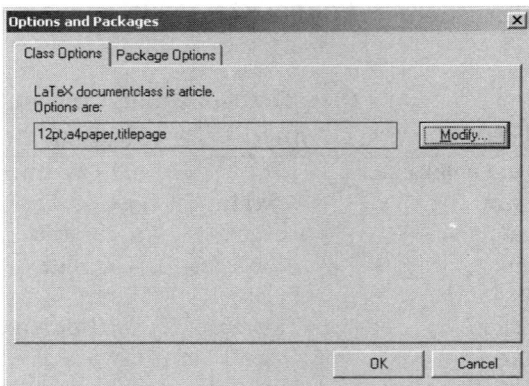

   In this example, three options use settings different from the defaults: body text font size (12pt), paper size (a4paper), and title page (titlepage).

4. Choose **OK** to return to your document.

Default Page Layouts

The following page layout images for the LaTeX article, book, and report document classes reflect the standard defaults for those classes. The keyed notes provide information about the size of the margins, headers, footers, text area, and margin notes, if any. Most measurements are given in points; a point is $\frac{1}{72}$ inch.

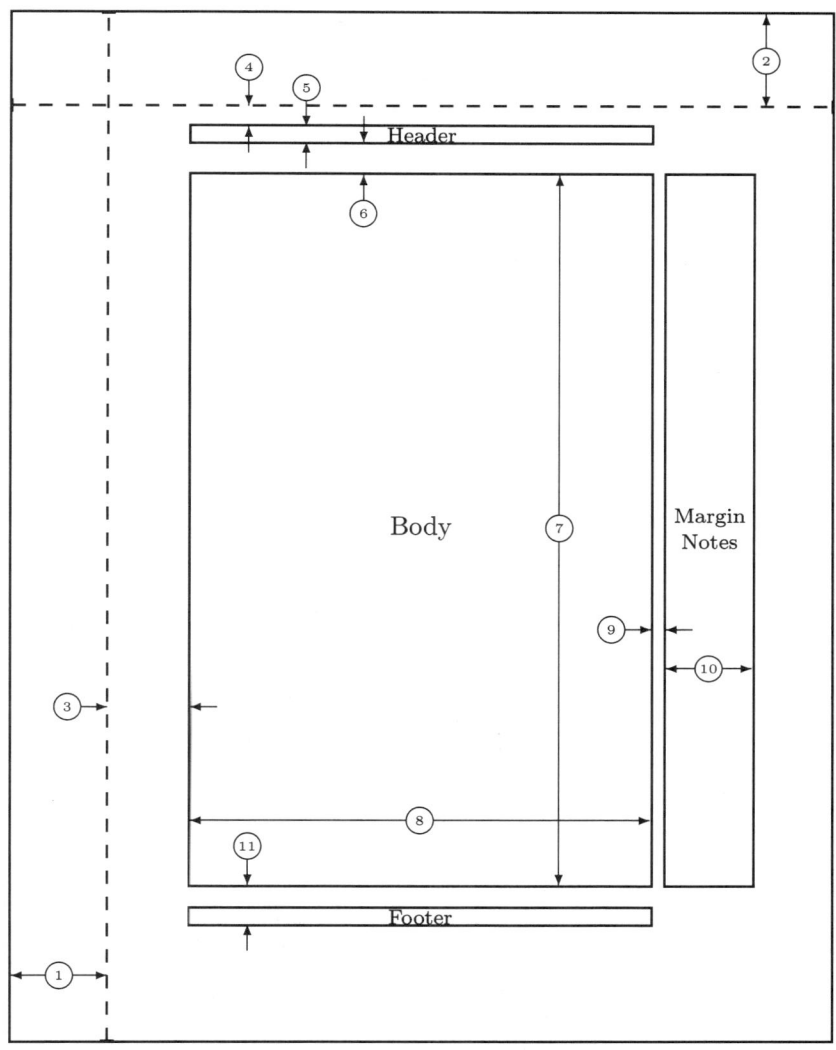

```
1   one inch + \hoffset         2   one inch + \voffset
3   \oddsidemargin = 62pt       4   \topmargin = 16pt
5   \headheight = 12pt          6   \headsep = 25pt
7   \textheight = 550pt         8   \textwidth = 345pt
9   \marginparsep = 11pt       10   \marginparwidth = 65pt
11  \footskip = 30pt                \marginparpush = 5pt (not shown)
    \hoffset = 0pt                  \voffset = 0pt
    \paperwidth = 614pt             \paperheight = 794pt
```

Default page layout for article.cls

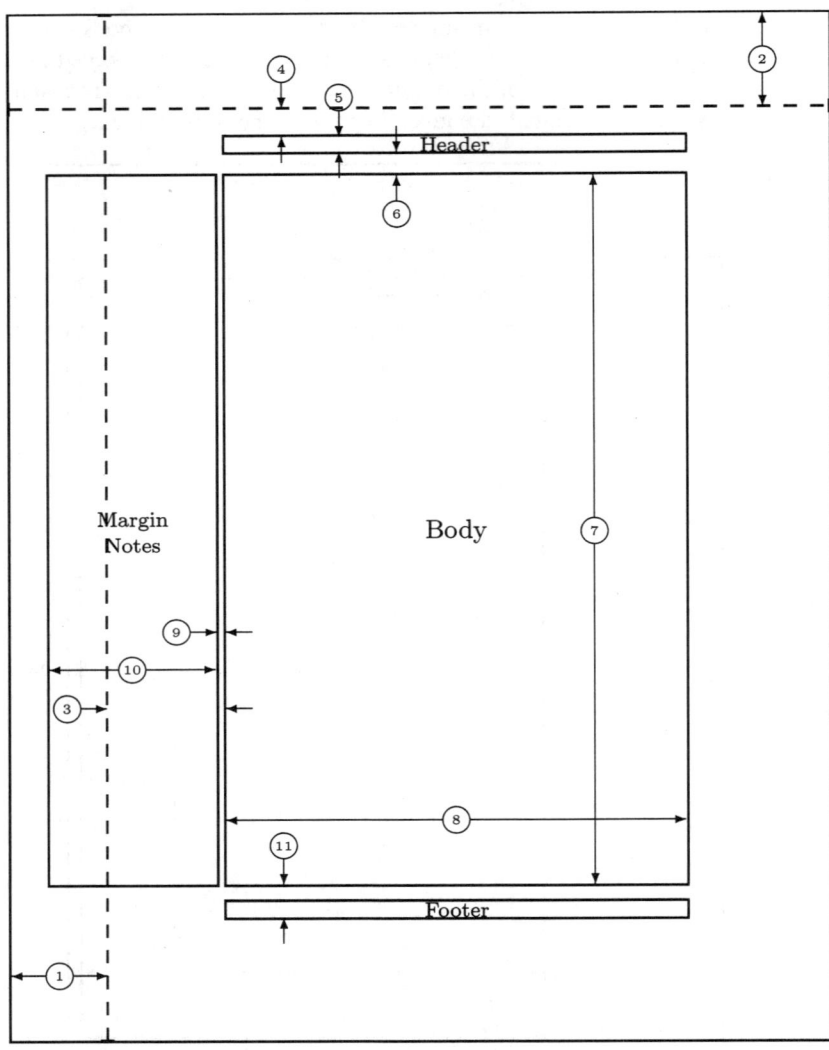

1	one inch + \hoffset	2	one inch + \voffset
3	\evensidemargin = 89pt	4	\topmargin = 22pt
5	\headheight = 12pt	6	\headsep = 18pt
7	\textheight = 550pt	8	\textwidth = 345pt
9	\marginparsep = 7pt	10	\marginparwidth = 125pt
11	\footskip = 25pt		\marginparpush = 5pt (not shown)
	\hoffset = 0pt		\voffset = 0pt
	\paperwidth = 614pt		\paperheight = 794pt

Default page layout for book.cls

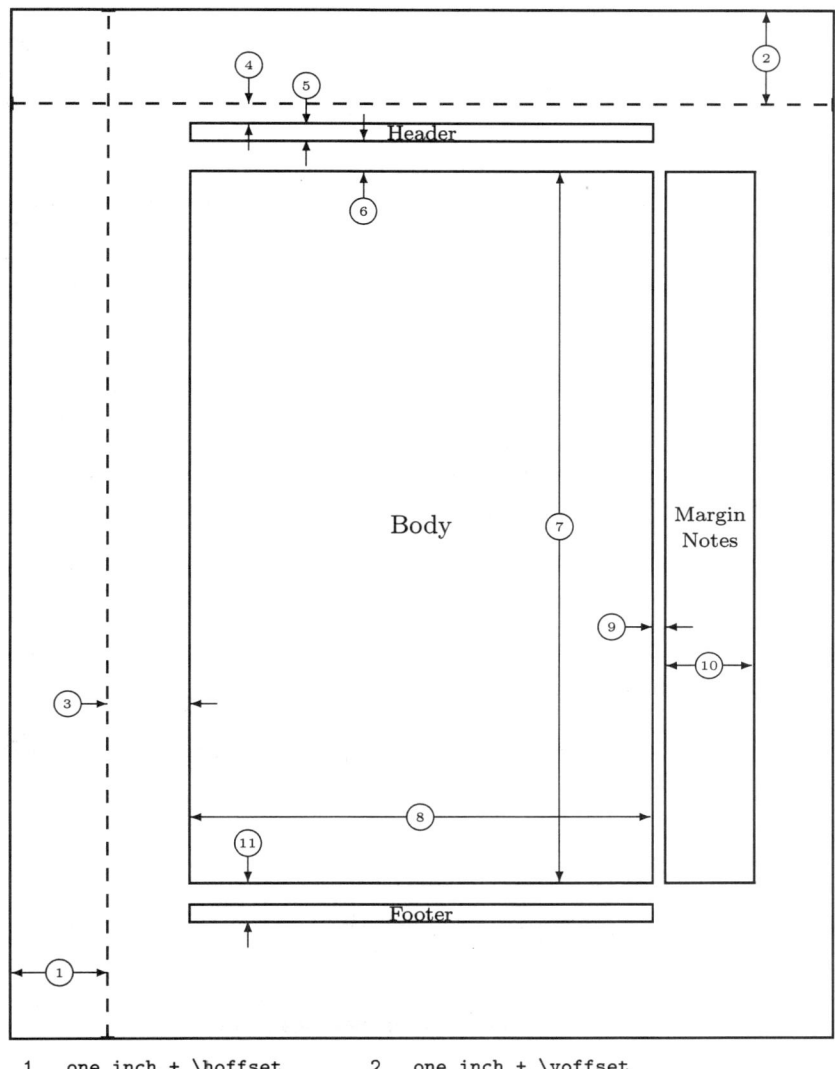

1	one inch + \hoffset	2	one inch + \voffset
3	\oddsidemargin = 62pt	4	\topmargin = 16pt
5	\headheight = 12pt	6	\headsep = 25pt
7	\textheight = 550pt	8	\textwidth = 345pt
9	\marginparsep = 11pt	10	\marginparwidth = 65pt
11	\footskip = 30pt		\marginparpush = 5pt (not shown)
	\hoffset = 0pt		\voffset = 0pt
	\paperwidth = 614pt		\paperheight = 794pt

Default page layout for report.cls

If your document uses different settings, these page layouts may not apply. For example, if a document uses a4 paper instead of 8.5x11 or two columns instead of one, the margins differ from those shown in these diagrams. You may want to add the *layout* package to your document to generate a page layout diagram (see page 116).

LaTeX Packages

The document class specifications establish a basic set of LaTeX typesetting instructions. LaTeX packages—sets of additional typesetting instructions—extend typesetting instructions by enabling some specific LaTeX behavior or customizing some aspect of the document appearance.

When you install *SWP* or *SW*, you automatically install those LaTeX packages that are included with the standard LaTeX distribution. These packages are installed in the `base`, `required`, and `AMS` subdirectories of the `TCITeX\TeX\LaTeX` directory. Additionally, the `TCITeX\TeX\LaTeX\contrib` directory of your installation includes a collection of CTAN packages, and the `TCITeX\TeX\LaTeX\SWmisc` directory contains packages from other sources including publishers and universities. Most packages have an `.sty` file extension.

Together, the packages enable a variety of customized typesetting behaviors, such as the creation of an index, the special formatting of footnotes, the content and design of headers and footers, the style of numbered lists, the generation of a list of symbols, and many others. Chapter 3 "Using LaTeX Packages" describes the packages available with *SWP* and *SW*. You can find links to additional and, often, extensive information about the packages in the online Help system.

When you open a document with a particular shell, the program automatically adds to the document any packages that are specified for the shell. You can easily determine which packages are in use for your document.

▶ **To determine the packages in use**

1. On the Typeset toolbar, choose the Options and Packages button ▦ or, from the **Typeset** menu, choose **Options and Packages**.

2. Choose the **Package Options** tab.

   The **Packages in Use** box lists the packages currently in use, as in this example:

   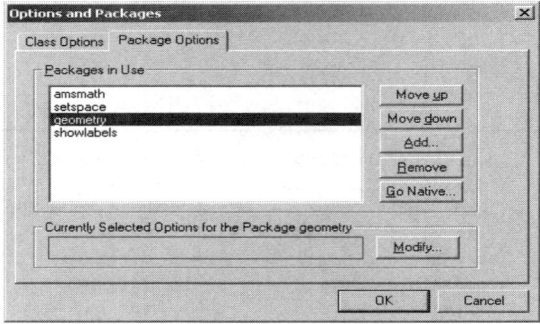

3. Choose **OK**.

   When you typeset your document, the program may call additional packages. For example, it adds the *amsmath* package by default to nearly every *SWP* and *SW* document. Also, many packages in turn call other packages when LaTeX or PDFLaTeX runs; that is, when you typeset the document. The list you see in the **Packages in Use** box includes only those packages that are directly called by the document.

You can add packages to your document in addition to those packages that have been specified for the shell. Documents created with most document shells—that is, documents in most document classes—can accept additional packages. The order in which the packages are specified can, on occasion, affect typesetting behavior. The documentation notes when package order is significant.

You can also remove a package from your document if you don't need the capability the package provides. See page 65 for information about adding and removing packages.

Note Unless you're very familiar with LaTeX and LaTeX packages, we urge you not to remove packages specified by the document shell.

Just as document classes have options, many packages have options with which you can customize document behavior still further. The options differ from package to package. Instructions for modifying packages begin on page 67.

When you typeset, LaTeX and PDFLaTeX use the class and package specifications to determine the typeset appearance of your document. Document classes and packages can conflict. If you expect to work with LaTeX packages, we urge you to learn how each package interacts with other packages, with document classes, and with the program. See also Chapter 3 "Using LaTeX Packages" for package information.

Choosing a Document Shell

Every *SWP* and *SW* document begins with a document shell. As you choose a shell for your new document, keep your typesetting needs in mind, especially if you expect to create a complex document.

Your installation includes an extensive collection of shells for creating new documents. In addition to many general purpose shells, we include with the program a collection of shells designed to meet the specific typesetting requirements of specific universities and scholarly journals. The shells produce documents that fall into several categories, which are reflected in the shell directories of your installation:

- **Articles**—short documents intended for publication in scholarly journals or conference proceedings.

- **Author Packages for AMS**—articles intended for publication in journals or conference proceedings published by the American Mathematical Society (AMS).

- **Books**—large documents intended for publication as a separate volume.

- **Exams and Syllabi**—short documents intended for use in the classroom.

- **International**—non-English documents, including German, Russian, Greek, Chinese, and Japanese.

- **Other Documents**—miscellaneous document types including faxes, letters, memos, overhead transparencies, slides, and some books and reports, usually developed for earlier releases of *SWP* and *SW*.

- **Scientific Notebook**—documents created with *Scientific Notebook*. Documents created with these shells are intended for printing without the benefit of typesetting.

- **Standard LaTeX**—documents created with the LaTeX base document classes without the addition of any packages.

- **Style Editor**—documents created with shells developed using the Style Editor.
- **Theses**—documents that fulfill thesis formatting requirements at several universities.

In addition to using the shells provided with the program, you can create your own shells; see page 70.

Make sure the shell you choose produces the type of document you want to create. Don't attempt to write a book using a letter shell or an article using a report shell. Make sure that the shell contains the tags appropriate for your work. If you need theorem environments, for example, choose a shell that has theorem and theorem-like item tags.

If you're unsure of your typesetting requirements, we urge you to choose the Standard LaTeX shell for the type of document you need. These standard shells provide the greatest flexibility and portability. You can achieve almost any typesetting effect by beginning with a standard shell and adding LaTeX packages as necessary.

Important We strongly recommend that you begin all new documents using one of the standard LaTeX shells, unless you have a compelling reason (such as publisher's instructions) to do otherwise.

A Gallery of Document Shells, provided on your program CD as a PDF file, illustrates the appearance of sample documents that have been typeset with each shell provided with the program. Examine the samples and note the features they illustrate, such as the absence or presence of headers and footers, the placement of page numbers and footnotes, the size of the margins, the appearance and placement of the headings, the extent of the front matter, the use of single or double columns, and the use of single or double spacing. When you find a shell that looks appropriate, open and print a new document with the shell to see if it meets your requirements. The closer the shell fits your requirements, the easier your typesetting tasks will be.

Each time you start *SWP* or *SW*, the program automatically opens a new, untitled start-up document using a default shell. If the shell is appropriate for your work, you can start entering information right away. If you want to create some other kind of document, however, open a new document with a different shell. You can change the default document shell to suit your needs.

▶ **To open a new document with the default shell**

- Start *SWP* or *SW*.
  The program automatically opens a new, empty document.

▶ **To open a new document with a different shell**

1. On the Standard toolbar, click the New button ▢ or, from the **File** menu, choose **New** to open the **New** dialog box.

2. From the **Shell Directories** list in the **New** dialog box, select the kind of document you want.

3. From the **Shell Files** list, select the shell you want and choose **OK**.

If most of the documents you create are similar, you can save time by changing the default shell so that the program automatically opens a start-up document that fits your needs.

▶ **To identify the default shell for start-up documents**

1. On the Editing toolbar, click [icon] or, from the Tools menu, choose User Setup, and then choose the Start-up Document tab.

   The default document shell is highlighted.

2. Choose OK.

▶ **To change the default shell**

1. On the Editing toolbar, click [icon] or, from the Tools menu, choose User Setup, and then choose the Start-up Document tab.

2. From the Shell Directories list, select the type of document you want.

3. From the Shell Files list, select the shell you want as the default and choose OK.

Tailoring a Shell to Your Needs

Once you've opened a new document with a shell that comes close to meeting your typesetting requirements, name and save the new .tex document. Then, you can begin tailoring the document to meet your requirements more precisely. When the document has the typeset appearance you want, save it (or in Version 4.0 and later, export it) as a new shell, as explained on page 70.

Although it is possible to achieve the typesetting results you want by tailoring your document outside *SWP* or *SW*, we focus here on tailoring your document *from within the program*. Several techniques for modifying the typesetting specifications are available. From within the program you can

- Modify the document class options.
- Add or remove LaTeX packages.
- Modify the LaTeX package options.
- Add TeX or LaTeX commands to the preamble or body of your document.

These techniques often involve adding raw TeX or LaTeX code to your document. That is, they involve adding code that isn't processed by the program but is rather passed directly to LaTeX or PDFLaTeX from *SWP* or *SW* when you typeset your document. Thus, you must be careful to enter commands using correct syntax to prevent LaTeX errors. Incorrect TeX or LaTeX code can cause permanent damage to your document. We strongly encourage you to save a copy of your document before you attempt any of the modifications suggested here.

Important Be sure to enter TeX or LaTeX commands correctly. Otherwise, you can damage your document permanently.

Modifying the Document Class Options

You can override the class option defaults set by the shell. As noted in Chapter 1 "Tailoring Typesetting to Your Needs," modifying the class options may be the easiest way to make the typesetting changes you need. The modification process is fast and easily reversed if it doesn't have the effect you want. If the shell you choose produces the typesetting results you want except, perhaps, for the body text font size, the size of the paper, or the default language, try modifying the class options before you attempt more complex modifications. We urge you not to modify the shell itself, but rather to save any modifications in a new shell, as explained on page 70.

If no options are listed when you try to modify the document class, you can *go native*, or add LaTeX commands to force the program to use a given option. When you typeset your document, the program passes the information directly to LaTeX for processing. If you enter incorrect commands, LaTeX may not be able to typeset your document and you can damage your document beyond repair.

Note Be careful to enter commands correctly. Incorrect commands can cause LaTeX to fail and may damage your document permanently.

▶ **To modify the class options**

1. On the Typeset toolbar, click the Options and Packages button or, from the Typeset menu, choose Options and Packages.

2. Choose the Class Options tab.

3. Choose Modify.

4. In the Category box, scroll the list to select the category you want to modify.

5. In the Options box, select the option you want and then choose OK.

6. If you want to add class options that aren't listed, choose Go Native, enter the command for the option you want, and choose OK.

7. Choose OK to return to your document.

8. Save your document and typeset preview it.

LaTeX interprets the class options according to the typesetting instructions in the `.cls` file for the document class. Some class options can take precedence over other instructions, such as those specified in packages; other class options may be ignored when certain packages are in use. In other words, although you may specify certain class option settings, LaTeX may ignore them. Document classes and packages don't always interact smoothly. If you're making extensive modifications, you may find that you must proceed by trial and error as you experiment with the various sets of specifications and learn how they interact.

Adding and Removing LaTeX Packages

The packages used by *SWP* and *SW* document shells have been carefully chosen to achieve certain typesetting results. However, you may decide that your document needs a package that has not yet been added. You may also decide that your document doesn't need a certain package; if so, you can remove it.

Note Unless you're very familiar with LaTeX and LaTeX packages, we urge you not to remove packages specified by the document shell.

From the **Package Options** tab, you can see the list of packages in effect and those that are available to be added to your document. Note that the list of available packages differs for different versions of the program. Remember that the order in which packages are specified can affect typesetting behavior.

On occasion, you may want to use a package that isn't part of your program installation. In that case, you must add it to the installation before you can add it to your document. See "Using Shells and Typesetting Specifications from Outside Sources" on page 71.

By default, the program automatically manages LaTeX packages, adding certain packages such as *amsmath* to most *SWP* and *SW* documents. If you have Version 5, you can manage packages yourself and prevent the program from adding packages automatically.

▶ **To add a package to your document**

1. On the Typeset toolbar, click the Options and Packages button 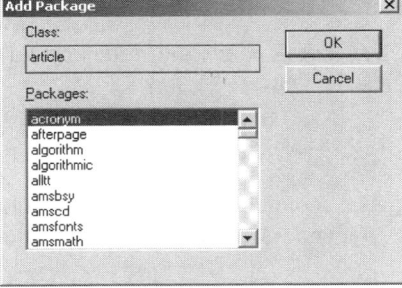 or, from the **Typeset** menu, choose **Options and Packages**.

2. Choose the **Package Options** tab.

3. If the package you want to add isn't listed in the **Packages in Use** box, choose **Add**.

4. In the **Add Packages** box, scroll the **Packages** list to select the package you want and then choose **OK**.

5. If you need to reorder the packages in the **Packages in Use** list, select a package and use the **Move Up** or **Move Down** controls to place the package in the correct position.

6. Choose **OK** again to return to your document.

If the package you want isn't listed as available in your version of *SWP* or *SW*, you can go native to add the LaTeX commands that force the program to use the package. When you typeset your document, the program passes the typesetting information directly to LaTeX for processing. If the commands are in error, LaTeX won't be able to typeset your document or to create a DVI or PDF file. Further, incorrect syntax can damage your document beyond repair.

Note Be careful to enter commands correctly. Incorrect syntax can cause LaTeX to fail and may damage your document permanently.

▶ **To add a package by going native**

1. On the Typeset toolbar, click the Options and Packages button or, from the Typeset menu, choose **Options and Packages**.

2. Choose the **Package Options** tab and choose **Go Native**.

3. Click the mouse in the **Native LaTeX Packages** dialog and scroll to the end of any commands that appear there.

4. Enter the name of the package you want, enclosed in curly braces.

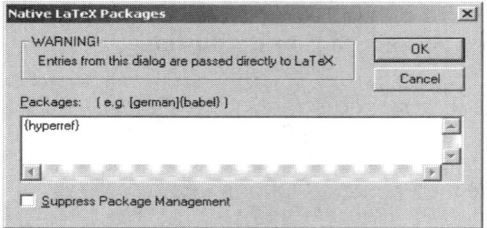

Remember: The program passes your entries directly to LaTeX and PDFLaTeX. Incorrect syntax will cause typesetting to fail.

5. Choose **OK** twice to return to your document.

You can add package options and arguments for the packages you specify. See "Modifying LaTeX Package Options" below.

▶ **To remove a package from your document**

1. On the Typeset toolbar, click the Options and Packages button or, from the Typeset menu, choose **Options and Packages**.

2. Choose the **Package Options** tab.

3. From the list of packages in the **Packages in Use** box, select the package you want to remove.

4. Choose **Remove**.

5. Choose **OK**.

▶ **To suppress program management of LaTeX packages**

1. On the Typeset toolbar, click the Options and Packages button ▦ or, from the **Typeset** menu, choose **Options and Packages**.

2. Choose the **Package Options** tab.

3. Choose **Go Native**.

4. Check **Suppress Package Management**.

5. Choose **OK** to close the dialog boxes and return to your document.

Modifying LaTeX Package Options

Many packages have a series of options for which you can specify settings, much as you specify class option settings. (To learn more the options available for specific packages, see Chapter 3 "Using LaTeX Packages" and the package documentation provided with the program.)

If a package has options available, the program generally lists them on the **Package Options** tab. If an option is marked as *default*, it is in effect. Defaults usually don't appear in the **Currently Selected Options** box. Selecting a default option has no effect other than to display it in the box.

When you select a package option setting that is listed in the **Package Options** tab, the program creates the correct LaTeX syntax for your selection. You can also go native to specify package option settings. However, if you go native, the program passes your commands directly to LaTeX without checking for correct syntax. Be careful to enter the commands correctly.

▶ **To modify package options**

1. On the Typeset toolbar, click the Options and Packages button ▦ or, from the **Typeset** menu, choose **Options and Packages** and then choose the **Package Options** tab.

2. Select the package you want to modify and choose **Modify**.

3. If the **Options** dialog lists options for the package,

   a. In the **Category** box, select the option you want.
   b. In the **Options** box, select the setting you want.
      The program displays the options you select in the **Currently Selected Options** area.

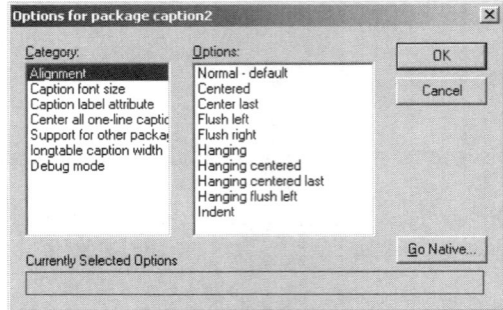

Repeat steps a and b for each option you want to modify.

c. Choose **OK** to return to the **Package Options** tab.

Note that the **Package Options** tab reflects the options you have selected.

or

If the **Options** dialog lists no options for the package,

a. Choose **Go Native**.

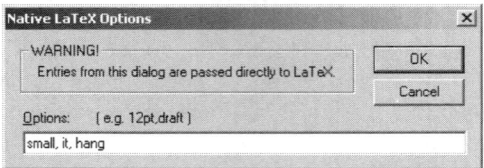

b. In the **Native LaTeX Options** dialog, enter the commands for any package options you want to apply.

The documentation for each package outlines the syntax and arguments of the commands for the available options. Additional information is available in Chapter 3 "Using LaTeX Packages." Remember that incorrect syntax can prevent typesetting and can damage your document.

c. Choose **OK** twice to return to the **Package Options** tab.

Note that the dialog reflects the options you have selected.

4. Choose **OK** to return to your document.

Adding TeX and LaTeX Commands to a Document

In addition to package options, many packages define a series of commands that you can use to send precise typesetting instructions to LaTeX or PDFLaTeX. The documentation for each package outlines the syntax and arguments of any available commands and explains whether the commands should be placed in the preamble or in TeX fields in the body of your document. In either case, you must make certain you enter correctly formatted LaTeX to avoid typesetting difficulties and possible damage to your document.

Tailoring a Shell to Your Needs   69

When you save your document, the program places any commands in the preamble before the \begin{document} statement. The preamble can contain definitions such as \newtheorem, \newcommand, \def, and \renewcommand, but must not contain any commands that generate typeset output. If you incorrectly enter any commands in the preamble, you can damage your document irreparably.

▶ **To add a command in the preamble of your document**

1. From the Typeset menu, choose Preamble.

2. If you're using Version 4.1 or earlier, click the mouse in the entry area.

   Caution If you begin typing without first clicking in the area, you will overwrite any commands already in the preamble. Choose Cancel to leave the preamble unchanged.

3. Enter the commands and choose OK.

▶ **To enter a TeX field**

1. If you have Version 4.0 or later, click the T_EX button on the Typeset Object toolbar or, from the Insert menu, choose Typeset Object and then choose TeX Field.

   If you have an earlier version, click the T_EX button on the Typeset Object toolbar or, from the Insert menu, choose Field and then choose TeX.

2. In the TeX Field dialog, type the T_EX command preceded by a backslash (\).

3. Choose OK.

The program inserts in your document a field containing the command. On the screen, the field appears as a small gray box containing the words *TeX field,* like this: TeX field. When you save your document, the program interprets the T_EX command and inserts it directly into the document file.

If you want to prevent the program from interpreting the command in a T_EX field, you can *encapsulate* and name the field. Then, when you save your document, the program stores the name and the command exactly as you entered them. The program preserves the exact syntax of the field when you save and reload the document. When you open the document again, the program displays the field on the screen as a small gray box with the field name in brackets, like this [create object]. When you typeset the document, T_EX interprets the encapsulated field and inserts the command in the DVI file.

Note Incorrect code in an encapsulated field won't cause the program to fail when you open your document, because the code remains hidden. However, it will prevent L^AT_EX from typesetting the document.

▶ To enter an encapsulated TeX field

1. If you have Version 4.0 or later, click the TeX button on the Typeset Object toolbar or, from the Insert menu, choose Typeset Object and then choose TeX Field.

   If you have an earlier version, click the TeX button on the Typeset Object toolbar or, from the Insert menu, choose Field and then choose TeX.

2. Check the Encapsulated box and enter a name for the field.

3. In the entry area, type the content and choose OK.

Creating Shells

You can create your own shells by saving any document as a shell file in one of the `Shells` subdirectories in your program installation. If you have carefully tailored a document so that its typeset output meets your needs, we urge you to save your work so that you can use it again. Similarly, if you have obtained typesetting specifications or shells from another source, such as a publisher, you should save them for future use. See Using Shells and Typesetting Specifications from Outside Sources on page 71.

Of course, you can use any *SWP* or *SW* document as the shell for a new document—it doesn't have to have an `.shl` extension—but creating a new shell removes the risk of changing something unintentionally in the original document. Further, if you place a new shell in one of the `Shell` subdirectories, its name appears in the shell list displayed when you start a new document. If you place the shell in some other directory, the name doesn't appear in the list of available shells and you can't create a new document with the shell using the New command. You can create new shell subdirectories as necessary.

In Version 4.0 and later, you save shells by exporting them as `.shl` files. Earlier versions use the Save As command.

▶ To create a shell

1. Open the document you want to use as a shell.

2. If you're using Version 4.0 or later, from the File menu, choose Export Document.

   or

   If you're using Version 3.5 or earlier, from the File menu, choose Save As.

3. Select a location for the new shell:

   a. In the box labeled Save in, select the `Shells` directory in your program installation.

   b. Select the appropriate subdirectory or create a new subdirectory for the new shell.

4. In the box labeled File name, type a name for the shell.

   The name can include spaces and nonalphabetic characters.

5. In the box labeled **Save as type**, specify **Shell (*.shl)**.

6. Choose **Save**.

   The next time you open a new document, the shell name appears on the **Shell Files** list corresponding to the shell subdirectory you specified.

Using Shells and Typesetting Specifications from Outside Sources

Although many shells and typesetting specifications are provided with *SWP* and *SW*, you may need to add specifications that you obtain from a publisher or from some other source for TeX and LaTeX files, such as the Comprehensive TeX Archive Network (CTAN). The CTAN directory on your program CD contains the typesetting specifications and files as distributed on CTAN, but only those files needed for typesetting are installed with the program.

We have tested the shells and specifications that we provide with the program to ensure that they work correctly and are compatible with *SWP* and *SW*. However, we can't guarantee that other specifications will work with our products, nor can we guarantee that the documents you create with those specifications will behave as the specifications advertise. Nonetheless, it's important to install specifications correctly, and we offer the instructions below as an aid.

Important We don't support documents created with typesetting specifications not provided with *SWP* or *SW*.

Adding LaTeX typesetting specifications to your installation involves these steps:

1. Placing the specification files in an appropriate directory.

2. Completing any required installation steps.

3. Testing the specifications by running the associated sample documents through LaTeX.

4. Opening the sample documents in *SWP* or *SW*.

5. Creating or modifying a `.cst` file, if necessary.

6. Creating a shell for the typesetting specifications.

Note Don't attempt to add LaTeX typesetting specifications to your installation if you aren't familiar with TeX and LaTeX.

We illustrate these instructions by showing how to install the specifications contained in `\CTAN\macros\latex\contrib\supported\uaclasses` on your program CD. These files adhere to the typesetting requirements for theses and dissertations at the University of Arizona.

These instructions assume you have installed *SWP* in the directory `c:\swp50`. If you're using *SW* or your directory is different, remember to substitute the correct directory path in the instructions.

Step One: Place the typesetting specification files in an appropriate directory

A set of LaTeX specifications usually involves a collection of files with extensions including .cls, .clo, and .sty. The specifications may also include installation files with an .ins extension. Most specifications have accompanying sample documents and readme files, which often contain installation instructions.

Whether you download the specifications from a website or receive them by email or on diskette, you must place them in the correct program subdirectory to ensure they are available to TrueTeX. You can place the files anywhere in the TCITeX\TeX directory or its subdirectories. We suggest that for each set of specifications you create a new subdirectory within TCITeX\TeX. If you obtained the files from CTAN, follow the directory structure used there.

Before you move the new specifications to a directory, search *SWP* or *SW* to make certain an older version of the specifications is not installed. If you find an older version, rename it before you add the new files to the installation directory structure.

▶ **To place typesetting specification files in a program directory**

1. Create a new subdirectory for the specifications within TCITeX\TeX.

   For our example, we create a new subdirectory called uaclasses in the existing directory c:\swp50\TCITeX\tex\latex\contrib\supported.

2. Move all the typesetting specification files to the new directory.

Step Two: Complete any required installation steps

Simply moving the files to the new directory may not complete the installation of the specifications. More steps may be required.

▶ **To complete the installation of typesetting specifications**

1. Read the readme file accompanying the specifications and follow any installation instructions it contains.

2. If you must process any files through LaTeX, use the TrueTeX Formatter outside *SWP* or *SW*:

   a. From the Windows **Start** menu, select *SWP* or *SW* and then select **TrueTeX Formatter**.
   b. In the **Open TeX File** dialog, specify the directory containing the specification files.
   c. In the **File name** box, type *.* to display all files in the directory.
   d. Select the file you want to process and choose **OK**.

   Processing these files often creates additional files required by the specifications. In our example, the readme file tells us to run LaTeX on two files: ua-classes.dtx and ua-classes.ins.

Step Three: Test the specifications by opening the associated sample documents

Most specifications have associated sample documents that demonstrate the features of the typesetting specifications. Test the sample documents by running them through LaTeX before you try to open them in *SWP* or *SW*.

▶ To test the sample documents

1. If a sample document is provided as a .tex file, process it through the TrueTeX Formatter outside *SWP* or *SW* as described in Step Two to create a DVI file.

   In our example, both `ua-example.tex` and `ua-example.dvi` are available. To make sure the installation is working, we copy and rename `ua-example.tex` as `newua-example.tex`, and then process the renamed file through the TrueTeX Formatter to create a new DVI file.

2. Preview the DVI file with the TrueTeX previewer:

   a. From the Windows **Start** menu, choose **Programs**.
   b. Select the *SWP* or *SW* submenu from the Windows **Programs** list and then select **TrueTeX Previewer**.
   c. From the **File** menu, choose **Open**.
   d. In the **Open DVI File** dialog, specify the directory containing the specification files.
   e. Select the file you want to process and choose **OK**.

Step Four: Open the sample documents in SWP or SW

Opening sample documents created with native LaTeX can have unpredictable results in *SWP* and *SW*. The program may not handle the documents correctly and might even crash while it tries to load the file. We suggest you make a copy of the .tex files for the sample documents before you attempt to open them.

▶ To open a sample document in SWP or SW

1. Make a copy of the .tex file if you have not already done so.

2. Choose [icon] or, from the **File** menu, choose **Open**.

3. Select the subdirectory you created in Step One.

4. Select the copy of the sample document and choose **OK**.

5. If the program displays a message indicating that an appropriate .cst file was not found, choose **Yes** to load the document using a default .cst file.

   The program chooses a default style from the appropriate `Styles` directory.

6. Typeset preview the document and compare the results to those obtained when you previewed the DVI file in Step Three.

   Similar results indicate a successful installation. If the document fails to compile, refer to the instructions on page 48 for finding and correcting LaTeX errors.

Step Five: Create and modify a new .cst file

When you open a document, the program uses the associated .cst file to display the document on the screen and to reflect the available environments, objects, and tags.

The .cst file has no effect on the document's typeset appearance. Because the new specifications aren't yet associated with a .cst file, you must create a new .cst file.

If the LaTeX specifications you're adding are similar to an existing document class and include no new objects, you can probably create a successful .cst file by copying and modifying the .cst file for a similar document class. The .cst files are installed in the Styles directory of your program installation or in one of its subdirectories. However, if the specifications you're adding implement a new base document class, you need to create a new .cst file that reflects all the environments in the new specifications. If the .cst file doesn't reflect all the environments in the new specifications, *SWP* or *SW* can't open the sample document successfully. After you have created the new .cst file, you must change the sample document to reflect the new .cst file.

In our example, the ua-classes specifications represent a new base document class called ua-thesis, so we must create a new .cst file and save it in a new subdirectory in the Styles directory. Then we have to modify the file so that the screen display reflects any new document elements, objects, and environments implemented by the typesetting specifications.

Modifying the .cst file involves determining which new objects are implemented by the typesetting specifications and then creating corresponding sections in the .cst file. Study the new typesetting specifications and the sample document, if any, to determine which tags the new .cst file must have. Look in the .cls file for new environments, theorem objects, and especially front matter elements. These objects are often signalled with \def or \newcommand statements. You may be able to find another .cst file that reflects the object. If so, you can copy the object to the new .cst file.

In our example, the ua-thesis specifications add an abstract to the standard LaTeX report class, so the .cst file must reflect the new object. We can search other .cst files to find an abstract object, copy it, and add it to ua-thesis.cst. You can find the resulting .cst file, Styles\ua-thesis\ua-thesis.cst, on your program CD.

▶ **To create a new .cst file**

1. Use an ASCII editor to open the .cls file associated with the new typesetting specifications.

2. Search for a statement that indicates the base document class for the new specifications.

   In our example, the .cls file indicates that the ua-thesis document class derives from the standard LaTeX report document class.

3. In the Styles directory of your program installation, find a .cst file that corresponds to the document class basis.

   In our example, we use report.cst in the Styles\report directory.

4. Rename the file using the same name as the document class name and save it in the Styles directory, either in a new subdirectory or in the [Special] subdirectory.

   We copy Styles\report\report.cst and rename the copy as Styles\ua-thesis\ua-thesis.cst.

5. If the new specifications have an environment that isn't contained in the .cst file, search for another .cst file that contains a similar object and copy the object to the new .cst file.

6. When you have added all necessary environments, save the new .cst file in a new subdirectory in the Styles directory.

   Note If the new subdirectory contains only one .cst file, you don't have to change the name of the .cst file in the sample document.

▶ **To change the sample document so that it will use the new .cst file**

1. Change the name of the .cst file:

   a. Open the file in *SWP* or *SW,* and from the **File** menu, choose **Style**.
   b. Choose **Advanced**.
   c. In the **Style File** box, browse to the directory containing the new .cst file.
   d. Select the file and choose **OK**.

2. Change the appearance of the tag environments in the document window, if necessary:
   - If you're using Version 4.0 or later, use the **Appearance** command on the **Tag** menu to change the tag attributes.
   or
   - If you're using an earlier version, use an ASCII editor to modify the tag attributes.

   Remember Modifications to the .cst file have no effect on the typeset appearance of your document.

Step Six: Create a new shell document for the typesetting specifications

We suggest you use the sample document as the shell document, modifying it as needed.

▶ **To create a shell document for the typesetting specifications**

1. In *SWP* or *SW,* open the .tex file for the sample document.

2. Make any changes you want to the file.

3. Save the file as a shell:

   a. If you're using Version 4.0 or later, from the **File** menu, choose **Export Document**.
   or
   If you're using an earlier version, from the **File** menu, choose **Save As**.
   b. In the **Save in** box, specify the directory for the shell.
   Choose an appropriate subdirectory within the Shells directory of your program installation. In our example, we save the shell as Thesis - University of Arizona Thesis.shl in the Shells\Theses directory.
   c. In the **File name** box, type the name of the shell.
   d. In the **Save as type** box, select **Shell (*.shl)**.
   e. Choose **Save**.

3 Using LaTeX Packages

LaTeX packages extend TeX typesetting capabilities by enabling some specific behavior for your document. The creation of an index, the inclusion of special bibliography lists, the use of color, the formatting of footnotes, and many other typesetting behaviors can be enabled with packages.

When you install *SWP* or *SW*, you automatically install those packages that are included with the standard LaTeX distribution. In addition, the installation includes packages that add specific typesetting capabilities to *SWP* and *SW* documents or that support the shells designed for certain publishers or universities. Certain packages are included in the installation only for purposes of compatibility with earlier versions of the program. Packages have an `.sty` extension and are loaded into the `TCITeX/TeX` directory and its various subdirectories at installation. The directory assignments, which are noted in the discussion below, reflect TeX convention:

Directory	Contents
`TCITeX/TeX/generic`	Input files used by many different formats
`TCITeX/TeX/LaTeX`	Files used with new versions of LaTeX
`TCITeX/TeX/latex209`	Files used only with LaTeX2.09*
`TCITeX/TeX/plain`	Files used only with Plain TeX

*Note that packages that are installed in the latex209 directory may or may not work for later implementations of LaTeX.

If the package you want doesn't appear in the list of packages available for your document, you can go native to add it, as described on page 66.

Most of the packages included with your installation work successfully with most *SWP* and *SW* documents; that is, you can correctly compile most *SWP* and *SW* documents to which one or more of these packages have been added, whether you are creating a device independent (DVI) file or, in Version 5, a Portable Document Format (PDF) file. However, even though LaTeX correctly compiles a document, you may not be able to preview it. Also, certain packages require the use of different print drivers. You may need to change your driver configuration to use certain packages.

Note, though, that when you change drivers, you tell LaTeX not to use the default driver configuration for the local LaTeX installation. If you subsequently try to compile your document in a different LaTeX installation, LaTeX will ignore the defaults for the new installation. Thus, you may need to make additional changes to your document to accommodate the new LaTeX setting. For the greatest portability, we recommend that you usually leave the driver configurations unchanged.

This chapter focuses on those packages that enhance general typesetting capabilities in most typeset documents. We have omitted a discussion of any packages (such as those whose names begin with `sw20`) that have been designed to support a single document shell rather than to provide a capability for LaTeX documents in general. You can learn more about these packages from their `.sty` files and often from the corresponding document shell. See also *A Gallery of Document Shells* on your program CD to explore the typeset appearance of documents created with the many shells provided with the program.

Organized alphabetically by package, this chapter explains the function of each package and briefly describes any available package options and commands. The chapter notes any known package conflicts with document classes and with preview and print drivers. Because complete instructions for using each package are outside the scope of this chapter, we encourage you to read the documentation accompanying the packages you want to use. You can find links to additional and, often, extensive package information in the `SWSamples\OptionsPackagesLaTeX.tex` file in your *SWP* or *SW* installation. Also, you may find helpful information in the `.sty` files for certain packages. A basic knowledge of TeX and LaTeX will help you understand some of the more technical information. See Chapter 2 "Using Document Shells" for information about basic program tasks related to using packages, such as adding packages, selecting options, and inserting commands in encapsulated TeX fields or in the document preamble.

Important Modifying the typesetting specifications can damage your document. Do not attempt extensive modifications unless you are familiar with TeX and LaTeX.

Although many packages are available for use with *SWP* and *SW* documents, you may be able to obtain all the typesetting capability you need by learning to use just a few of them. The table beginning on page 43 will help you identify the packages that you need most often. In particular, these packages often prove useful: *breakcites*, *caption2*, *cite*, *color*, *endnotes*, *fancyhdr*, *float*, *geometry*, *longtable*, *nomencl*, *setspace*, *tocbibind*, and *wrapfig*.

Acronym

The *acronym* package helps you manage acronyms and acronym lists in your documents. You can define each acronym within a special acronym environment and then use macros in the text to define how each occurrence of the acronym will appear when you typeset the document. If you define the list in the document preamble, it appears before the body of the document. If you define the list in the body of the document, it appears where you place it. You may want to add a heading to designate the list. The *acronym* package requires that you typeset your document with two LaTeX passes for proper resolution of any acronyms in use.

The program doesn't understand the macros used by the package, but you can successfully use the macros in your document if you place them inside encapsulated TeX fields.

▶ **To define acronyms in the text**

1. Add the *acronym* package to your document.

2. Begin the acronym environment:

   a. Place the insertion point where you want the list to appear in your document.
   b. Enter a TeX field.

c. In the entry area, type \begin{acronym}.

d. Choose OK.

3. For each acronym,

   a. Enter an encapsulated TeX field.

   b. To define the acronym and include it in the list of acronyms, type
   \acro{*acronym*}{*definition*} and choose OK.
   or
   To define the acronym and exclude it from the list of acronyms, type
   \acrodef{*acronym*}{*definition*} and choose OK.

4. Following the last definition, end the environment:

   a. Enter an encapsulated TeX field.

   b. In the entry area, type \end{acronym} and choose OK.

▶ **To define acronyms in the preamble**

1. Add the *acronym* package to your document.

2. From the Typeset menu, choose Preamble.

3. Click the mouse in the entry area.

4. On a new line, type \begin{acronym} and press ENTER.

5. For each acronym,

   - To define the acronym and include it in the list of acronyms, type
     \acro{*acronym*}{*definition*} and press ENTER.
     or
   - To define the acronym and exclude it from the list of acronyms, type
     \acrodef{*acronym*}{*definition*} and press ENTER.

6. Type \end{acronym} and choose OK.

▶ **To use acronyms**

1. Place the insertion point where you want an acronym to appear.

2. Enter an encapsulated TeX field.

3. In the entry area, type the command to insert the acronym formatted according to your preferences:

Command	Effect
\ac{*acronym*}	Expand and identify the acronym the first time; use only the acronym thereafter
\acf{*acronym*}	Use the full name of the acronym
\acs{*acronym*}	Use the acronym, even before the first corresponding \ac command
\acl{*acronym*}	Expand the acronym without using the acronym itself

Suppose you've defined the acronym *SW* as *Scientific Word*. Now you want to use it in the sentence *(acronym) documents are beautifully typeset*.

These examples show the result of using the four available acronym commands, assuming that the acronym has already been used once in the document:

Command	Effect
\ac{*SW*}	SW documents are beautifully typeset.
\acf{*SW*}	Scientific Word (SW) documents are beautifully typeset.
\acs{*SW*}	SW documents are beautifully typeset.
\acl{*SW*}	Scientific Word documents are beautifully typeset.

4. Choose OK.

In addition to using the available commands, you can change the package option to place expanded acronyms in the body of the document or at the foot of the page as footnotes. The option is available through the **Options and Packages** command on the **Typeset** menu.

See an example of the package in use in the `PackageSample-acronym.tex` file in the `SWSamples` directory of your program installation. The package is installed in the `TCITeX/TeX/LaTeX/contrib/supported/acronym` directory.

Afterpage

The package implements the \afterpage command and causes LaTeX to expand its argument after the current page is filled and output. Although you can specify any command in the \afterpage argument, using the \clearpage command is a particularly useful way to force the printing of any floating objects (graphics and long tables) that haven't yet been anchored to a position. LaTeX fills the page on which the \afterpage command occurs and then prints any unanchored floating objects before continuing with the text. Use the \afterpage command in an encapsulated TeX field. The package has no options.

▶ **To use the afterpage package to output floating objects**

1. Add the *afterpage* package to your document.

2. Place the insertion point on the page after which you want accumulated floating objects to appear.

3. Enter an encapsulated TeX field.

4. In the entry area, type **\afterpage{\clearpage}**.

5. Choose OK.

Although you can add the *afterpage* package to documents in most document classes, note that the package doesn't work for two-column layouts. The package is installed in the `TCITeX/TeX/LaTeX/required/tools` directory and is part of the Standard LaTeX Tools Bundle.

Algorithm

The package provides two environments for describing algorithms—*algorithmic* and *algorithm*. The environments are designed to be used together, but they can be used separately.

The algorithmic environment provides an environment and commands for describing complex algorithms. Available commands include if-then-else constructs; for, while, until, and infinite loops; pre- and postconditions; and comments. Please see the package documentation for information about the commands. Line numbering is optional. An option to print end statements is available for the algorithmic environment.

The algorithm environment enables algorithm numbering and provides a floating environment for algorithm descriptions so they don't break over a page boundary. If you're using a report or book shell, you can produce a list of numbered algorithms for inclusion after the table of contents. The list of algorithms appears on a separate page, similar to a list of figures or list of tables. You must process your document through LaTeX outside *SWP* or *SW* to generate the list. Run LaTeX at least twice.

Options that affect the appearance and numbering of algorithm environments are available through the **Options and Packages** command on the **Typeset** menu.

▶ **To use the algorithm environments**

1. Add the *algorithm* package to your document.

2. If you want to use both algorithmic environments, add the *algorithmic* package.

3. If it is present, remove the definition of the algorithm newtheorem environment from the document preamble:

    a. From the **Typeset** menu, choose **Preamble**.
    b. Click the mouse in the entry area.
    c. Delete the definition, which looks something like this:
       \newtheorem{algorithm}[theorem]{Algorithm}
    d. Choose **OK**.

4. Enter the algorithm:

    a. Enter an encapsulated TeX field.
    b. Type **begin{algorithm}** or **begin{algorithmic}**.
    c. Enter the commands for the entire algorithm.
    d. To end the algorithmic environment, type **end{algorithmic}** or **end{algorithm}**.
    e. Choose **OK**.

▶ **To add a list of algorithms**

1. Place the insertion point at the beginning of the body of your document.

2. Enter an encapsulated TeX field.

3. Type **listofalgorithms** and choose **OK**.

4. Save the document.

5. From outside *SWP* or *SW,* typeset compile the document file:

   a. From the *SWP* or *SW* program group, choose the TrueTeX Formatter.
   b. Select the file and choose **OK**.
      LaTeX generates an .loa file for the document.

6. Typeset preview the document.

The packages are installed in the TCITeX/TeX/LaTeX/contrib/supported/algorithms directory. See an example of the package in use in the PackageSample-algorithm.rap file in the SWSamples directory of your program installation.

Alltt

The *alltt* package provides a verbatim-like environment in which the meaning of slashes and curly braces is unchanged by LaTeX. Thus, you can embed other TeX commands and environments inside the alltt environment to produce formatted mathematics and mathematics symbols.

Assume you want to include mathematics in a typeset verbatim paragraph. Ordinarily, the mathematics appears as LaTeX code when you typeset the document without the *alltt* package, as shown here:

```
This verbatim paragraph contains both text
and mathematics. Here, the use of the
Pythagorean theorem, $a\sp{2}+b\sp{2}=c\sp{2}$,
demonstrates this feature.
```

With the package, the mathematics appears as correctly formatted mathematics:

```
This verbatim paragraph contains both text
and mathematics. Here, the use of the
Pythagorean theorem, a² + b² = c²;
demonstrates this feature.
```

You must place the entire alltt environment in an encapsulated TeX field.

▶ **To use mathematics in a verbatim-like environment**

1. Add the *alltt* package to your document.

2. Enter an encapsulated TeX field.

3. Type **\begin{alltt}** to begin the alltt environment.

4. Begin entering the content of the verbatim environment.

5. For each mathematical element,

a. Type \(to begin mathematics.
   b. Type the commands for the mathematical statement or symbol you want.
   c. Type \) to end mathematics.

6. Complete the content.

7. Type **\end{alltt}** to close the environment.

8. Choose OK.

The package has no options. *Alltt* is provided automatically with LaTeX and is installed in the `TCITeX/TeX/LaTeX/base` directory.

AMS Packages

The American Mathematical Society ($\mathcal{AMS}$) publishes three main types of publications: articles, proceedings, and books or monographs. Each has detailed publication format specifications, which are reflected in three $\mathcal{AMS}$ shell documents: AMS Journal Article; AMS Proceedings Article; and AMS Book or Monograph. The specifications are supported by $\mathcal{AMS}$-LaTeX, a required component of the standard LaTeX distribution, and by a series of $\mathcal{AMS}$ packages. Most of the packages are installed in the `amscls` and `amsmath` subdirectories of the `TCITeX/TeX/LaTeX` directory. The *amsfonts* package is installed in the `TCITeX/TeX/plain` directory.

AMSFonts
The *amsfonts* package is a collection of fonts of symbols and characters that aren't always included in standard distributions of TeX, but that correspond to those used in $\mathcal{AMS}$ print and online publications and in the MathSci online database. The fonts include Blackboard Bold, Fraktur, the Euler family; certain sizes of Computer Modern mathematics, caps, and small caps fonts; extra mathematical symbols; and Cyrillic. Many *SWP* and *SW* document shells automatically call the amsfonts package.

Other than adding the package to your document, no action is required. No options are available for the *amsfonts* package. Note that the program adds the package if it is needed by features in use in the document. The Portable LaTeX filter always adds the *amsfonts* package, along with *amsmath, amssymb,* and *graphicx*.

AMSSymb
The *amssymb* package is a subset of amsfonts that defines the full set of symbol names for two fonts of extra symbols included in the amsfonts collection. The two fonts, msam and msbm, contain symbols, including uppercase Blackboard Bold, needed by the $\mathcal{AMS}$ publishing program and MathSci online database. The package requires no special commands in the document, and no options are available. The program adds the package if it is needed by features in use in the document.

AMSMath
This package, which is provided automatically with LaTeX, enhances the typeset appearance of mathematical formulas, especially those involving displayed equations, multi-line sub- and superscripts, and other mathematical constructs. The *amsmath* package

is included automatically in most SWP and SW shells. The program adds the package if it is needed by features in use in the document.

The package calls several auxiliary packages as needed:

- *amstext*—Allows typesetting of a small amount of text inside a display through the use of a TeX command.
- *amsopn*—Allows the declaration of new operator names.
- *amsbsy*—Included for backward compatibility only. This package has been superseded by the newer *bm* package that comes with LaTeX.

You can use two other packages in documents created with $\mathcal{AMS}$ document shells:

- *amscd*—Provides a CD environment for commutative diagrams; doesn't support diagonal arrows.
- *amsxtra*—Provides miscellaneous elements that enhance compatibility with documents created using earlier versions.

AMSMath Options

With the **Options and Packages** command on the **Typeset** menu, you can set six options for the *amsmath* package. They affect the placement of limits, equation numbers, and equations themselves. The options you set for this package may override options set for the document class.

Answers

The *answers* package provides a way to bind a solution to an exercise in a LaTeX environment. This package was designed for the general LaTeX community and may not be the best choice for SWP and SW documents. We urge the use of the Exam Builder.

With the *answers* package, you can store the bound solutions in several different files at once, so that you can print them at different times, such as in the appendix of a book as well as a handout for students. Further, you can create and include many different solutions files in a document, such as one for each section or chapter of a book. The package supports any number of solution types, including hints for students.

Available commands associate the exercises with the solutions and define, open, and close the solutions files. See the package documentation for more information about using the commands and for examples of using the package. The option to create solutions files is available through the **Options and Packages** command on the **Typeset** menu.

The package is installed in the `TCITeX/TeX/LaTeX/contrib/supported/answers` directory.

Apacite

The *apacite* package formats citations according to the complex requirements of the American Psychological Association (APA). The package works with the bibliography style file `apacite.bst` to produce citations in a variety of APA formats. It improves on the capabilities of the *apalike, apalike-plus,* and *newapa* packages. In particular, the package provides "no parentheses" citation commands. Except in rare cases, the package will format every reference correctly.

Commands are available to handle various types of citations. See the package documentation for more information about using the package. The *apacite* package is installed in `TCITeX/TeX/latex/contrib/other/bibtex`.

Apalike and Apalike-plus

The *apalike* package formats text according to specifications in the American Psychological Association *Publication Manual (4th edition),* to produce typesetting suitable for APA journals. In particular, the package works in conjunction with `apalike.bst` to produce BibTeX bibliography entries that are formatted alphabetically by author's last name. The package also produces single and multiple author-date citations in the text.

The *apalike-plus* package extends the features of *apalike* with TeX commands that provide optional titles for the list of references and include the selected title in headers and the table of contents. The commands are as follows:

- \bibtitle—generates *References* as the default bibliography title.
- \bibheadtitle—generates *REFERENCES* as the default text to be used in page headers.
- \addcontentsline{toc}{...}{\bibtitle}—generates an appropriately titled entry for the bibliography in the table of contents.

These two packages have no effect on manual bibliographies. To use the packages successfully, you must specify that you want to create a BibTeX bibliography and choose the corresponding BibTeX style.

▶ **To create a list of references with apalike or apalike-plus**

1. Add the *apalike* or *apalike-plus* package to your document.

2. From the **Typeset** menu, choose **Bibliography Choice**.

3. Check **BibTeX** and choose **OK**.

4. Insert BibTeX citations as needed throughout your document.

5. Insert the list of references:

   a. Place the insertion point where you want the bibliography to appear in your document.
   b. If you're using *apalike-plus,* enter package commands in an encapsulated TeX field to specify the bibliography title and page headers and to include the list in the table of contents.
   c. From the **Insert** menu, choose **Typeset Object** and then choose **Bibliography**.
   d. Select the BibTeX database file you want to use.
   e. Scroll down the **Style** list to select **apalike.bst** or **apalike2.bst**.
   f. Choose **OK**.

6. Save and compile the document.

No options are available for either package. Both *apalike* and *apalike-plus* are installed in `TCITeX/TeX/latex209/contrib/misc`.

Appendix

The *appendix* package provides for modifying the typesetting of appendix titles. It provides a subappendix environment for use as an appendix to a chapter or section. Although package commands are available, you can use the package more effectively with the options available from the **Options and Packages** command on the **Typeset** menu. The options affect the formatting of headers and titles in the appendix.

The subappendices environment creates an appendix section at the end of a chapter or an appendix subsection at the end of a section. It numbers the subappendix in sequence with the other sections or subsections and attaches an uppercase letter to the section number. Subappendices appear in the table of contents.

▶ **To create a subappendix**

1. Add the *appendix* package to your document.

2. Begin the subappendices environment:

    a. Place the insertion point where you want the subappendix to appear.
    b. Enter an encapsulated TeX field.
    c. In the entry area, type **begin{subappendices}** and choose **OK**.

3. Type a heading for the subappendix.

4. If the subappendix is in a chapter, apply the section tag to the heading.

    or

    If the subappendix is in a section, apply the subsection tag to the heading.

5. Enter the content of the subappendix.

6. End the subappendices environment:

    a. At the end of the subappendix, enter an encapsulated TeX field.
    b. In the entry area, type **end{subappendices}** and choose **OK**.

The *appendix* package is designed to work only with those document classes that have chapters or sections. The package is known to conflict with the LaTeX kernel \include command. See the package documentation for additional information. The *appendix* package is installed in `TCITeX/TeX/LaTeX/contrib/supported/appendix`.

Array

The *array* package extends the implementation of the LaTeX array and tabular environments by providing options for column formatting, including lines and paragraph indention. You can use the package to achieve alignment within cells, like this:

You can also obtain special effect using vertical rules with variable widths:

and you can format paragraph indention within cells:

The package provides other column spacing capabilities.

The package has no options, so you must enter commands for the entire tabular environment in an encapsulated TeX field. See the package documentation for instructions and for additional examples of package effects. The package is installed in the `TCITeX/TeX/LaTeX/required/tools` directory and is part of the Standard LaTeX Tools Bundle.

Astron

The *astron* package produces author-year citations in two forms: (Author, year) and (year). It is required by these BIBTeX bibliography styles:

- `astron.bst`—produces bibliographies in the format required by the European astronomical journal *Astronomy and Astrophysics*.
- `apa.bst`—produces bibliographies in the American Psychological Association format.
- `bbs.bst`—produces bibliographies approximately in the format of *Behavioral and Brain Sciences*.
- `cbe.bst`—produces bibliographies approximately in the Council of Biology Editors format.
- `humanbio.bst`—produces bibliographies with a format similar to that used in *Human Biology*.

- `humannat.bst`—produces bibliographies with a format of *Human Nature and American Anthropologist*.
- `jtb.bst.`—produces bibliographies based loosely on the format used in the *Journal of Theoretical Biology*.

Other than adding the package to your document, no action is required. The *astron* package is installed in the `TCITeX/TeX/LaTeX/contrib/other/bibtex` directory.

Authordate1-4

The package implements four options for creating author-date citations. The options are required when using the BIBTeX styles `authordate1.bst`, `authordate2.bst`, `authordate3.bst`, or `authordate4.bst`. The package has no effect on manual bibliographies. It produces BIBTeX bibliographies in four slightly differing formats:

- **authordate1** produces author-date reference lists with the author's name typeset in Roman. Any uppercase letters that occur in the titles of articles, journals, or books are left as given in the BIBTeX file.
- **authordate2** produces author-date reference lists with the author's name typeset in Roman and "downstyle" titles. That is, working from the BIBTeX file, the package changes to lowercase any uppercase letters except the first that occur in the titles of articles, journals, or books; any letter that follows a colon; and any letters protected by the right and left parenthesis marks.
- **authordate3** produces author-date reference lists with the author's name typeset in small capitals. Otherwise, the lists are as produced by authordate1.
- **authordate4** produces author-date reference lists with the author's name typeset in small capitals and downstyle titles as in authordate2.

The formats are based loosely on the recommendation of *British Standard 1629 (1976 edition)*, Butcher's *Copy-editing* (Cambridge University Press, 1981) and *The Chicago Manual of Style (1982 edition)*.

Be sure to select BIBTeX bibliographies from the **Bibliography Choice** dialog on the **Typeset** menu. Once you have added the package to your document, you must specify the .bst file you want when you insert the BIBTeX field.

▶ **To create an author-date reference list**

1. Add the *authordate1-4* package to your document.

2. From the **Typeset** menu, choose **Bibliography Choice**.

3. Check **BibTeX** and choose **OK**.

4. Insert BIBTeX citations as needed throughout your document.

5. Specify a .bst file for the authordate package:

   a. Place the insertion point where you want the bibliography to appear in your document.

   b. From the **Insert** menu, choose **Typeset Object** and then choose **Bibliography**.

   c. Select the BIBTeX database file you want to use.

d. Scroll down the Style list to select the .bst file for authordate1, authordate2, authordate3, or authordate4.

   e. Choose OK.

6. Save and compile the document.

The package has no options. The package is installed in the `TCITeX/TeX/LaTeX/contrib/other/bibtex` directory. See the *harvard* package on page 111 and the *chicago* package on page 95 for information about other ways to create author-date citations.

Babel

The *babel* package addresses language-specific issues so that TeX works more reliably to typeset documents written in languages other than English. When the appropriate language hyphenation patterns are included in the format file, the package switches the active hyphenation patterns as the base language is switched. The multilingual format file created with a standard *SWP* or *SW* installation includes these hyphenation patterns: English, American English, French, German, and German new orthography.

If you need a different pattern, you must use a different TrueTeX format file. The program CD for Version 3.51 and for Version 4.1 and later (but not for Version 4.0) includes several format files in the `Extras/TrueTeX/TrueTeXFormatFile` directory. If these don't include the hyphenation pattern you need, you must create a format file that does. You can find instructions in the online Help.

The *babel* package also corrects problems with embedded English strings in LaTeX, such as *Chapter* or *Bibliography*. When *babel* is running with a specific language, it uses strings appropriate for that language in place of the embedded English strings. However, theorem objects must be treated separately. Typically, words that are typeset in the lead-in objects of theorem statements are set in the \newtheorem statements in the document preamble. To change the words, modify the statements in the document preamble. With the *babel* package, LaTeX can successfully typeset multiple languages in the same document. The language options shown on page 90 are available through the Options and Packages command on the Typeset menu.

Once you have added the package to a document, you can switch languages within your document. When you typeset the document, LaTeX uses the appropriate language to typeset embedded strings and hyphenate text.

▶ **To typeset documents using multiple languages**

1. Add the *babel* package to your document.

2. Ensure that the appropriate language hyphenation patterns are included in the format file in use.

3. Modify the package options to select the language or languages you want.

   LaTeX uses the last language you specify as the default language for embedded strings and hyphenation. See page 56 for information about viewing package options.

4. If you want to switch to a different language at some point in your document,
   a. Place the insertion point where you want to begin the language.
   b. Enter an encapsulated TeX field.
   c. In the entry area, type **\selectlanguage**{*language*}
      where *language* is the language you want to use at this point in the document. Be sure you have selected the language option.
   d. Choose **OK**.
   When you typeset, LaTeX treats the new language correctly.

No standard exists for transporting files that rely on the availability of a certain language. Each file must be handled on an ad hoc basis. In Version 3.5 and earlier, the package requires the Multilingual LaTeX installation option. The *babel* package is installed in the `TCITeX/TeX/LaTeX/required/babel` directory.

Language	Options	Language	Options
Afrikaans	Include, Exclude	Hungarian	Magyar, Hungarian, None of the above
Bahasa	Include, Exclude		
Breton	Include, Exclude	Irish Gaelic	Include, Exclude
Catalan	Include, Exclude	Italian	Include, Exclude
Croatian	Include, Exclude	Lower Sorbian	Include, Exclude
Czech	Include, Exclude	Norwegian	Norsk, Nynorsk, None of the above
Danish	Include, Exclude		
Dutch	Include, Exclude	Polish	Include, Exclude
English	English, U.S. English, American, UK British, British, None of the above	Portuguese	Portuges, Portuguese, Brazilian, Brazil, None of the above
		Romanian	Include, Exclude
Esperanto	Include, Exclude	Russian	Include, Exclude
Estonian	Include, Exclude	Scottish Gaelic	Include, Exclude
Finnish	Include, Exclude	Spanish	Include, Exclude
French	Use French, Use francais, None of the above	Slovakian	Include, Exclude
		Slovenian	Include, Exclude
Galician	Include, Exclude	Swedish	Include, Exclude
German	Austrian, Austrian new orthography, German, German new orthography, GermanB, None of the above	Turkish	Include, Exclude
		Ukrainian	Include, Exclude
		Upper Sorbian	Include, Exclude
		Welsh	Include, Exclude
Greek	Greek, Polutroniko, None of the above		

Bibmods

The *bibmods* package modifies the TeX thebibliography environment to improve spacing, especially for two-column documents. Adding the package to your document provides the package functions; no further action is required. No options or commands are defined for the package. The package is installed in the `TCITeX/TeX/latex209/contrib/misc` directory.

Blkarray

The *blkarray* package defines array and tabular environments not unlike those defined by the array package. When the insertion point is in math, the *blkarray* package implements a blockarray environment that functions similarly to the array environment in standard LaTeX. When the insertion point is in text, the package implements an environment that functions similarly to the tabular environment.

However, the *blkarray* package differs in that it defines column types differently, making all column specifiers equal. The package lends itself to detailed formatting of blockarray environments. The package implements different formatting for blocks of cells within a table, such as a header row. Information can span several columns and doesn't have to be aligned with information in other cells. You can add rules to separate rows.

Additionally, the package implements the use of delimiters as column specifiers. That is, you can use a delimiter around blocks of cells within an array, like this:

$$\begin{bmatrix} 1 & 2 \\ 4 & 5 \\ 7 & 8 \\ 1 & 2 \\ 4 & 5 \end{bmatrix} \begin{matrix} 3 \\ 6 \\ 9 \\ 3 \\ 6 \end{matrix} \Bigg\}$$

Blkarray environments can accept footnote commands. Depending on options selected, resulting footnotes may appear at the end of the table or the foot of the page.

Package options aren't available. You must specify them in encapsulated TeX fields, just as you do for the array package. The content of the encapsulated field includes and defines the entire tabular environment. See the package documentation for complete instructions and for additional examples of blkarray effects. The standard LaTeX command \hline doesn't work with *blkarray*. While the package produces various kinds of tables, it may not be an appropriate substitution for array and tabular environments. The package is installed in `TCITeX/TeX/LaTeX/contrib/supported/carlisle`.

Boxedminipage

The package creates a LaTeX minipage environment surrounded by rules, like this:

> This environment is useful for emphasizing information.

You can control the width of the environment. Additionally, you can use the standard TeX commands \fboxrule and \fboxsep to determine the thickness of the rules and the distance between the rules and the inside edge of the box, respectively.

> This environment is useful for emphasizing information of a mournful nature.

No package options are defined for the package. Instead, you enter the package commands in encapsulated TeX fields. The package is installed in `TCITeX/TeX/latex209/contrib/misc`.

▶ To use the boxedminipage environment

1. Add the *boxedminipage* package to your document.

2. Place the insertion point where you want the boxed environment to begin.

3. Enter an encapsulated TeX field and type **\begin{boxedminipage}{x}** where *x* is the desired width of the minipage.

   The command for first example on page 91 is \begin{boxedminipage}{1.75in}. The commands for the second are

   \setlength{\fboxrule}{4pt}
   \setlength{\fboxsep}{12pt}
   \begin{boxedminipage}{3in}

4. Choose **OK**.

5. Move the insertion point to the end of the information you want to box.

6. Enter an encapsulated TeX field, type **\end{boxedminipage}** and choose **OK**.

Breakcites

The *breakcites* package allows LaTeX to create line breaks within long citations or citations with remarks, resulting in better line spacing in your typeset document. No action is required beyond adding the package to your document and creating the citations you need. The package has no options. It is installed in the `TCITeX/TeX/LaTeX/contrib/other/misc` directory.

Caption and Caption2

The *caption* package and its newer beta version, *caption2*, implement customized captions within floating environments. In particular, the packages allow definition of the caption width and alignment, caption label font size and font attributes, and caption font size. The packages also support rotated captions for floating objects that are presented sideways.

Note that the TrueTeX Previewer provided with *SWP* and *SW* doesn't support rotation; you must use a different DVI previewer and print driver if you want to use the caption or caption2 package to rotate captions in a DVI file. However, PDF viewers do support rotation, so you can use the package to create rotated captions in typeset PDF files.

With the **Options and Packages** command on the **Typeset** menu, you can set options that affect the alignment, centering, font size, and font attributes of captions. No additional commands are required. The installation program places these two related packages in the `TCITeX/TeX/LaTeX/contrib/supported/caption` directory.

Chapterbib

With the *chapterbib* package, you can create a BIBTeX bibliography for each file you include in your document. If each included file represents a separate chapter, then each chapter can have its own bibliography. Thus, your document can contain multiple small BIBTeX bibliographies as well as a comprehensive BIBTeX bibliography for the whole. You can also create a bibliography for the whole document, such as a recommended reading list, that is unrelated to cited works. The bibliography items can be cited in more than one bibliography.

Your document requires several changes to use *chapterbib*. Each file that you include must have its own \bibliographystyle and \bibliography commands. To generate the bibliography, you must run BIBTeX on each included file separately. If you also want a bibliography for the whole document, the master document should have its own \bibliographystyle command. Generally, to generate the bibliographies, you must typeset your document (one pass through LaTeX), run BIBTeX on each included file, and then typeset your document again (two passes through LaTeX). The more complex your document, the more complex the process.

▶ **To generate bibliographies for included files**

1. Add the *chapterbib* package to your master document.

2. Scroll through the document to find each subdocument you have included.

    Subdocuments appear as gray boxes containing the subdocument name.

3. Replace each subdocument with a TeX field:

    a. Enter an encapsulated TeX field.
    b. In the entry area, type **include**{*subdoc*} where *subdoc* is the name of the subdocument to be included. Do not include the .tex file extension; LaTeX will provide that automatically.
    c. Choose **OK**.

4. Save the document.

5. From outside *SWP* or *SW,* typeset compile the document file:

    a. From the *SWP* or *SW* program group, choose the TrueTeX Formatter.
    b. Select the file and choose **OK**.
       LaTeX generates .aux files for each subdocument included.

6. Run BIBTeX on the .aux file for each subdocument and for the main document, if it also has a bibliography:
    - In Version 4.0 and later,
      i From the **Typeset** menu, choose **Tools**.
      ii Choose **Run BibTeX**.
      or

- In Version 3.5,
  i From the Windows menu, choose Run.
  ii In the Open box, type **swp35\TCITeX\SWTools\bin\bibtex.exe** and choose OK.
  Change the name of the program directory as necessary.
  iii Specify the .aux file.
  iv Choose Create.
  BIBTeX creates a .bbl file.
  v Choose OK.

7. From outside *SWP* or *SW*, typeset compile the document file at least twice more.

8. From inside *SWP* or *SW*, typeset preview the document.

Additional commands provide customized entries in a list of citations and multiple bibliographies without using the \include command. With package options you can repeat or gather all chapter bibliography entries at the end of the document, create a bibliography for the entire document, and format the bibliography title.

This package is compatible with three others: *cite* (see page 96), *overcite* (see page 124), and *drftcite* (see page 101), all installed in the TCITeX/TeX/LaTeX/contrib/supported/cite directory.

Chbibref

The LaTeX document class article sets a default name of *References* for the bibliography, but the report and book classes set the default name to *Bibliography*. The *chbibref* package sets a standard name for the bibliography for all three LaTeX document classes.

▶ **To change the title of the bibliography for all three document classes**

1. If you're using Version 3.5 or lower, obtain and install the *chbibref* package.

   The package is distributed with later versions.

2. Add the *chbibref* package to your document.

3. From the Typeset menu, choose Preamble, and click the mouse in the entry area.

4. At the end of the preamble, add a new line and type **\setbibref{*name*}** where *name* is the bibliography title you want.

   Note If you're using *babel,* place the command after the \begin{document} command.

5. Choose OK.

No package options are available. The package is installed in the TCITeX/TeX/LaTeX/contrib/supported/misc directory.

Chicago

The *chicago* package is used in combination with `chicago.bst` to produce BibTeX bibliographies formatted according to *The Chicago Manual of Style, Edition 13*.

▶ **To create a bibliography formatted according to** *The Chicago Manual of Style*

1. Add the *chicago* package to your document.

2. From the **Typeset** menu, choose **Bibliography Choice**.

3. Check **BibTeX** and choose **OK**.

4. Specify the `chicago.bst` file:

    a. Place the insertion point where you want the bibliography to appear.
    b. From the **Insert** menu, choose **Typeset Object** and then choose **Bibliography**.
    c. Select the BibTeX database files you want to use.
    d. Scroll down the **Style** list to select `chicago.bst` and choose **OK**.

5. Save and compile the document.

The package also supports a variety of citation formats. Although the program interface doesn't directly support these modifications, you can achieve the citation format you want by inserting commands in TeX fields.

Command	Citation Format
\cite{key}	Full author list and year: (Pearson 2002; Swanson, MacKendrick, and Medd 2000)
\citeNP{key}	Full author list and year, but without enclosing parentheses: Pearson 2002; Swanson, MacKendrick, and Medd 2000
\citeA{key}	Full author list without year: (Pearson; Swanson, MacKendrick, Medd)
\citeANP{key}	Full author list without parentheses: Pearson; Swanson, MacKendrick, Medd
\citeN{key}	Full author list, no parentheses around authors, parentheses around year: Swanson, MacKendrick, Medd (2000) note that....
\shortcite{key}	Abbreviated author list and year: (Swanson et al. 2000)
\shortciteNP{key}	Abbreviated author list and year, no parentheses: Swanson et al. 2000
\shortciteA{key}	Abbreviated author list: (Swanson et al.)
\shortciteANP{key}	Abbreviated author list, no parentheses: Swanson et al.
\shortciteN{key}	Abbreviated author list and year, parentheses around year: Swanson et al. (2000)
\citeyear{key}	Year information only, with parentheses: (2000)
\citeyearNP{key}	Year information only, without parentheses: 2000

▶ To modify the format of bibliography citations

1. Place the insertion point where you want the citation to appear.

2. Enter an encapsulated TeX field.

3. Type the command for the citation format you want, substituting the key for the BIBTeX reference.

4. Choose **OK**.

No package options are available. For more information, open a new document with the Standard LaTeX Article (Chicago) shell. The package is installed in the `TCITeX/TeX/LaTeX/contrib/other/bibtex` directory.

Cite

The *cite* package orders numerical citations and compresses a list of at least three consecutive numerical citations that occur together in the text. For example, a citation of [7, 5, 1, 4] becomes [1, 4, 5, 7] and a citation of [2, 6, 4, 7, 3] becomes [2–4, 6,7] with the use of the cite package.

To use the package, either use the citation command from the **Insert** menu or enclose your lists of citations in encapsulated TeX fields.

▶ To order and compress citations created with SWP and SW citations

1. Add the *cite* package to your document.

2. Place the insertion point where you want the citation to occur.

3. Choose [icon] or, from the **Insert** menu, choose **Typeset Object** and then choose **Citation**.

4. In the **Key** box, enter the key for the reference you want to cite or select the key from the dropdown list.

5. Select the contents of the **Key** box and copy the selection to the clipboard.

6. In the **Key** box, enter the key for the next reference.

    Note that because the first key was selected, the new key overwrites the first one.

7. Type a comma.

8. Paste the contents of the clipboard to the **Key** box.

9. Repeat steps 5–8 until you have entered all the references you need.

10. Choose **OK**.

When you typeset your document, LaTeX orders the list. Whenever the list includes three references in numerical sequence, LaTeX compresses them.

Command variations are available; see the package documentation for more information about using the package. In addition, several citation spacing options are available for the *cite* package through the **Options and Packages** command on the **Typeset** menu.

The *cite* package is one of four compatible packages installed in the `TCITeX/TeX/LaTeX/contrib/supported/cite` directory. The packages are *drftcite* (see page 101), *overcite* (see page 124), and *chapterbib* (see page 93).

Color

The *color* package produces boxes or entire pages with colored backgrounds. The package implements LaTeX support for color when the active typeset output driver can produce colored text. Many driver options are available through the **Options and Packages** command on the **Typeset** menu. For flexibility, we recommend that you leave the options unmodified. The local LaTeX installation sets the driver defaults. If you leave the configuration unchanged, you can compile your document without changes in another LaTeX environment. The package works successfully with PDF files.

▶ **To use color in a document**

1. Add the *color* package to your document.

2. If necessary, change the driver configuration:

   a. Modify the package options to select the driver you want to use.
   b. Use the **Expert Settings** command on the **Typeset** menu to modify the format, preview, and print driver settings to reflect the driver.
   For instructions, see the online Help or *Creating Documents with Scientific Workplace and Scientific Word*. The drivers must have been installed separately. They aren't included with *SWP* or *SW*.
   Important Don't attempt to modify the driver settings if you're not very familiar with TeX and LaTeX.

You can predefine the colors you want to use in the preamble of your document or specify them at the point you need them with commands in encapsulated TeX fields. The commands specify whether you want a box or a page in color. The commands also specify which of four common color models you want to use: rgb (red, green, blue); cmyk (cyan, magenta, yellow, black); gray; or named (names known to the selected driver). The monochrome option turns off all colors and is useful if you want to preview your document using a previewer that cannot produce color. Command arguments specify the exact color. See the package documentation for more information about using the package commands.

The *color* package is part of the Standard LaTeX Graphics Bundle along with *graphicx* (see page 111). For more information, see the `PackageSample-color.tex` file in the `SWSamples` directory of your program installation. The package is installed in the `TCITeX/TeX/LaTeX/required/graphics` directory.

Colortbl

The package produces colored background panels and rules for specified columns or rows of a table or array. The package implements LaTeX support for color when the active typeset output driver can produce colored text.

You can add color to a row in a table by inserting a package command in the table. Adding color to a column is more complex: you must enter the *colortbl* package commands, along with commands for the entire tabular environment, in an encapsulated TeX field. You indicate the size of each color panel and the corresponding color you want with commands placed at the start of the tabular environment.

The basic command syntax is as follows:

$$\text{\textbackslash columncolor}[w]\{x\}[y][z] \text{ or } \text{\textbackslash rowcolor}[w]\{x\}[y][z]$$

where

w is the color model: rgb (red, green, blue); cmyk (cyan, magenta, yellow, black); gray; or named (names known to the selected driver),

x is the selected color,

y is the amount of left overhang past the widest entry in the column, and

z is the amount of right overhang past the widest entry in the column.

See the package documentation for instructions and for additional examples of package effects.

▶ **To add color to a table row**

1. Add the *colortbl* package to your document.

2. Create a table.

3. Place the insertion point in the table at the beginning of the row you want to appear in color.

4. Enter an encapsulated TeX field.

5. In the entry area, type \rowcolor[w]{x}[y][z], completing the command as described above.

6. Choose OK.

▶ **To add color to a table column**

1. Place the insertion point where you want the table to appear.

2. Enter an encapsulated TeX field.

3. In the entry area, enter the complete tabular environment.

4. Place the insertion point at the beginning of the column you want to appear in color.

   For example, if your table is defined with the command \begin{tabular}{|l|l|l|} and you want to add color to the first column, place the insertion point after the first |.

5. Type >{\columncolor[*w*]{*x*}[*y*][*z*]}, completing the command as defined above.

6. Choose OK.

Many driver options are available through the Options and Packages command on the Typeset menu. The package is installed in the TCITeX/TeX/LaTeX/contrib/supported/carlisle directory. It should work successfully with other packages that have syntax compatible to that of the *array* package, such as *longtable* and *dcolumn*. The package works successfully with PDF files.

Comma

The *comma* package formats LaTeX counter values so that they print with a separator (such as a comma) every three digits. That is, the package prints a counter value of 1374 as 1,374 when the comma package is added to the document. The default separator is a comma, but you can customize it to use a period, thin space, or any other symbol that can be represented by a TeX command. LaTeX uses the same separator for all counters.

To use the package, you must add TeX commands to the document preamble or insert TeX fields containing those commands in the body of your document. To place the separator in all counters of more than three digits, place the command at the start of the body of the document. LaTeX applies the command from that point forward.

▶ **To insert separators in printed LaTeX counters**

1. Add the *comma* package to your document.

2. Add the package commands to the preamble of your document:

   a. From the Typeset menu, choose Preamble.
   b. Click the mouse in the entry area and scroll to the end of the entries.
   c. Type \renewcommand\the*counter*{\commaform{*counter*}} where *counter* is the LaTeX counter you want to print with a separator, such as the counter for section or theorem numbers.
   d. If you want to change the separator from a comma to some other symbol, add \renewcommand\commaformtoken{*x*} where *x* is any TeX command.
   e. Choose OK.

   or

   Add the commands in the body of the document:

   a. Place the insertion point at the start of the text.
   b. Enter an encapsulated TeX field.
   c. Type the package commands you need.
   d. Choose OK.

No package options are available. The package is installed in the TCITeX/TeX/LaTeX/contrib/supported/carlisle directory.

Dcolumn

The *dcolumn* package provides decimal point alignment for columns of entries in a tabular or array. You can define the separator (usually a period or comma) on which the columns align. Also, you can define a separator for the DVI output file and the maximum number of decimal places allowed in the column.

You must enter the *dcolumn* command, along with the commands for the entire tabular environment, in an encapsulated TeX field. The basic syntax of the column separator command is

D{*separator for the tex file*}{*separator for LaTeX*}{*decimal places*}

where *separator for the tex file* is the punctuation used in the document to indicate the decimal point; *separator for LaTeX* is the punctuation you want LaTeX to use when you typeset; and *decimal places* is the maximum number of decimal places in the column. A negative number in the *decimal places* argument indicates that any number of decimal places is acceptable.

▶ **To align column entries on a decimal point**

1. Place the insertion point where you want the table to appear.

2. Enter an encapsulated TeX field.

3. In the entry area, type **\newcolumntype{d}[0]{D**{*separator for the tex file*}{*separator for LaTeX*}{*decimal places*}**}**, completing the command as defined above.

4. Enter the complete tabular environment, beginning with a **\begin{tabular}{d....d}** command and ending with an **\end{tabular}** command.

5. Choose OK.

No package options are available. The package is installed in the `TCITeX/TeX/LaTeX/required/tools` directory and is part of the Standard LaTeX Tools Bundle.

Delarray

The package enhances the array package by adding a system of large paired delimiters around the array. This feature is built into *SWP* and *SW*.

▶ **To define delimiters for an array**

1. Add the *delarray* package to your document.

2. Place the insertion point where you want the delimited array to appear.

3. Enter an encapsulated TeX field.

4. Type **\[\begin{array}|{c}|** where | is the delimiter you want and *c* is repeated for each column in the array.

5. Enter the contents of the array.

6. Type **\end{array}\]** and choose OK.

The package has no options. The package is installed in the `TCITeX/TeX/LaTeX/required/tools` directory and is part of the Standard LaTeX Tools Bundle.

Doublespace

This package has been superseded by the *setspace* package (see page 135) and is included in *SWP* and *SW* for compatibility purposes. The package produces double spacing by redefining the LaTeX parameter \baselinestretch to 2. After you have added *doublespace* to your document, you need no additional commands to create a document that is double spaced throughout, with the exception of footnotes. No package options are available. The *doublespace* package is installed in the `TCITeX/TeX/latex209/contrib/misc` directory.

Drftcite

The *drftcite* package is designed to manage citations and bibliography items in document drafts. Remove the package or replace it with the *cite* or *overcite* package when you're ready for final printing. *Drftcite* forces LaTeX to print citations and reference lists using the labels of the bibliography items instead of their numbers. (LaTeX stores the correct citation numbers in the .aux file for subsequent use.) In the reference list, the package uses superscripted numbers to note the order in which bibliography items are cited in the text. Thus, the numbers also indicate the order in which the bibliography items should appear in the reference list. Uncited bibliography items are easy to find because they have no superscripted number. Unlike *cite* and *overcite*, the *drftcite* package doesn't order the items in the citation list. *Drftcite* works with both BibTeX and manual bibliographies. Create citations as usual from the Insert menu or Typeset Field toolbar.

▶ **To print citations and reference lists using item labels**

1. Add the *drftcite* package to your document.

2. Place the insertion point where you want a citation to occur.

3. Click [icon] or, from the Insert menu, choose Typeset Object and then choose Citation.

4. Enter the key of the reference you want to cite.

5. Choose OK.

Several citation formatting options are available through the Options and Packages command on the Typeset menu.

The *drftcite* package is one of four compatible packages—including *cite* (see page 96), *overcite* (see page 124), and *chapterbib* (see page 93)—installed in the `TCITeX/TeX/LaTeX/contrib/supported/cite` directory.

Dropping

The *dropping* package creates large dropped letters, as you see at the beginning of this paragraph. The number of lines over which you want the letters to extend determines their size. Experimentation will help you determine the most visually pleasing size for your document.

▶ To enter a dropped letter

1. Add the *dropping* package to your document.

2. Place the insertion point at the beginning of a paragraph.

3. Enter an encapsulated TeX field.

4. In the entry area, type **\dropping[x]{y}{z}** where x indicates how far from the left margin the letter should start (0pt is the default); y is the number of lines over which you want the capital letter to extend; and z is the letter or letters you want to enlarge.

5. Choose **OK**.

6. Type the remainder of the sentence.

In addition, many driver options are available through the **Options and Packages** command on the **Typeset** menu. We recommend that you leave the driver option unchanged. The package is in the `TCITeX/TeX/LaTeX/contrib/other/dropping` directory.

Endnotes

The package forces LaTeX to produce footnotes in a list of notes set in small type at the end of the document, instead of as footnotes set on the bottom of the page on which they occur. The package stores the endnotes in an extra external file with the file extension .ent. LaTeX generates a new version of the .ent file each time you typeset the document.

▶ To replace footnotes with endnotes

1. Add the *endnotes* package to your document.

2. Modify the document preamble:

    a. From the **Typeset** menu, choose **Preamble**.
    b. Click the mouse in the entry area.
    c. Scroll to the end of the entries and add a new line.
    d. Type **\let\footnote=\endnote** and choose **OK**.

3. Create the footnotes:

    a. From the **Insert** menu, choose **Note**.
    b. In the **Type of Note** box, select **footnote**.
    c. Type the footnote and choose **OK**.

4. Place the insertion point at the end of the document, where you want the endnotes to appear.

5. Enter an encapsulated TeX field.

6. Enter the commands you need:

   a. If you want the endnotes to begin on a new page, type **newpage** and press ENTER.
   b. Type **begingroup** and press ENTER.
   c. If you want an entry for the endnotes to appear in the table of contents, type **addcontentsline{toc}{section}{Notes}** and press ENTER.
   d. If you want the endnotes to be set in a normal size font instead of a smaller font, type **renewcommand{\enotesize}{\normalsize}** and press ENTER.
   e. Type **theendnotes** and press ENTER.
   f. Type **endgroup** and choose OK.

7. Typeset your document.

   LaTeX places any footnotes in the document in a list at the location of the encapsulated field.

The package implements additional commands to produce numbered endnotes; produce the endnote mark in the text but no corresponding endnote; or produce an endnote but no corresponding mark in the text. Additional commands change the endnote size and produce endnote numbers or marks. See the package documentation for more information about using the package and see an example of the package in use in the `PackageSample-endnotes.tex` file in the `SWSamples` directory of your program installation. No package options are available for the endnotes package. The package is installed in the `TCITeX/TeX/LaTeX/contrib/other/misc` directory.

Enumerate

The *enumerate* package provides an optional argument for the enumerate (numbered list) environment so that you can define the style in which LaTeX prints the counter. Argument parameters include designations for upper and lowercase alphabetic characters, upper and lowercase Roman numerals, and arabic numerals. To use the *enumerate* package, you must place the entire list in an encapsulated TeX field. Lists may be nested.

▶ **To define a counter style for numbered lists**

1. Add the *enumerate* package to your document.

2. Place the insertion point where you want the numbered list to appear.

3. Enter an encapsulated TeX field.

4. Type **begin{enumerate}[*x*]** where x is the designation of the counter numbering scheme:

Scheme	Example	Produces
A	A, B, C...	uppercase letters (as produced by \Alph)
a	a, b, c...	lowercase letters (as produced by \alph)
I	I, II, III...	uppercase roman numerals (as produced by \Roman)
i	i, ii, iii...	lowercase roman numerals (as produced by \roman)
1	1, 2, 3...	arabic numbers (as produced by \arabic)

5. Choose **OK**.

6. For each item in the list,

    a. Enter an encapsulated TeX field.
    b. Type **item** *text* where *text* is the content of the list item.
    c. Choose **OK**.

7. At the end of the list, enter an encapsulated TeX field.

8. Type **end{enumerate}**.

9. Choose **OK**.

No package options are available for the *enumerate* package. The package is installed in the `TCITeX/TeX/LaTeX/required/tools` directory as part of the Standard LaTeX Tools Bundle.

Euler

The *euler* package produces mathematics in LaTeX documents using the $\mathcal{AMS}$ Euler family of fonts (Euler Roman, Euler Fraktur, Euler Script, and Euler Extension). After you add the *euler* package to your document, LaTeX produces all mathematics in the document using the Euler font family. Several package options for formatting with Euler fonts are available through the **Options and Packages** command on the **Typeset** menu. The package is in the `TCITeX/TeX/LaTeX/contrib/supported/euler` directory.

Exscale

The *exscale* package provides different sizes of large mathematical symbols in LaTeX documents. The symbols are based on certain sizes of the cmex10 font and on cmex 7-point to 9-point variants, which are part of the $\mathcal{AMS}$ font package. The *exscale* package is redundant in *SWP* and *SW* documents, in which mathematical operators and delimiters are scaled automatically.

The package has no options. *Exscale* is installed in the `TCITeX/TeX/LaTeX/base` directory.

Fancybox

The *fancybox* package provides several different styles of boxes for framing and rotating content in your document. *Fancybox* provides commands that produce boxes with shadows, square-cornered boxes with single or double lines, and round-cornered boxes with normal or bold lines, such as these:

Text with shadow box   Text with oval box   Text with double box

The boxes can contain words, lines, paragraphs, or whole pages, and the boxes can be centered or right- or left-justified.

Note that the TrueTeX Previewer provided with *SWP* and *SW* doesn't support rotation; you must use a different DVI previewer and print driver if you want to rotate boxed content in a DVI file. However, PDF viewers support rotation, so you can use the package to create rotated boxes in typeset PDF files.

Package options aren't available for the *fancybox* package. Use encapsulated TeX fields to box information in your document.

▶ **To box information in your document**

1. Add the *fancybox* package to your document.

2. Enter an encapsulated TeX field.

3. Type ***command**{*content to be boxed*}* where ***command*** is the fancybox command you need:

Command	Effect
fbox	square box
shadowbox	square box with shadow
doublebox	double square box
ovalbox	thin oval box
Ovalbox	thick oval box

4. Choose OK.

Package documentation includes useful information about using LaTeX box macros. The `PackageSample-fancybox.tex` file in the `SWSamples` directory of your program installation contains more examples of fancybox effects. The package is installed in the `TCITeX/TeX/LaTeX/contrib/supported/fancybox` directory.

Fancyhdr

The *fancyhdr* package is a page layout customizing package that replaces the *fancyheadings* package. It handles footers and headers efficiently, but also works with placement of floats. With the package, you can define headers and footers with multiple parts and on multiple lines, place rules in headers and footers, and use a header and footer width different from that of the text. Additionally, you can use different headers and footers for even and odd pages, first pages of chapters, and on pages containing floats, and you can produce dictionary-style headers reflecting the first and last words on a page. *Fancyhdr* also provides control over fonts and upper- and lowercase letters.

No package options are available. However, the package uses a simplified syntax for entering commands.

▶ **To define the content of headers and footers**

1. Add the *fancyhdr* package to your document.

2. Specify a new header and footer setup in the document preamble:

    a. From the **Typeset** menu, choose **Preamble**.
    b. Click the mouse in the entry area.
    c. On a new line, type **\pagestyle{fancy}**.

3. Define the content of the header:

    a. On a new line, type **\lhead{*text*}** where *text* is the information you want left-justified in the header.
    b. On a new line, type **\chead{*text*}** where *text* is the information you want centered in the header.
    c. On a new line, type **\rhead{*text*}** where *text* is the information you want right-justified in the header.
    d. If you want a rule under the header, type **\renewcommand{\headrulewidth}{*x*}** where *x* is the point size of the rule you want.
    Note The *text* argument can contain TeX commands.

4. Define the content of the footer:

    a. On a new line, type **\lfoot{*text*}** where *text* is the information you want left-justified in the footer.
    b. On a new line, type **\cfoot{*text*}** where *text* is the information you want centered in the footer.
    c. On a new line, type **\rfoot{*text*}** where *text* is the information you want right-justified in the footer.
    d. If you want a rule over the footer, type **\renewcommand{\footrulewidth}{*x*}** where *x* is the point size of the rule you want.
    Note The *text* argument can contain TeX commands.

5. Choose **OK**.

Defining different headers and footers for double-sided documents increases the definitions, some for even (left) and others for odd (right) pages. You can occasionally combine definitions in the same command by specifying when they should appear. The package uses these settings:

Setting	Prints on
E	Even page
O	Odd page
L	Left
C	Center
R	Right
H	Header
F	Footer

▶ **To define the content of different headers and footers for even and odd pages**

1. Add the *fancyhdr* package to your document.

2. Specify a new header and footer setup in the document preamble:

    a. From the **Typeset** menu, choose **Preamble**.
    b. Click the mouse in the entry area.
    c. On a new line, type **\pagestyle{fancy}**.
    d. On a new line, type **\fancyhf{}** to clear all fields in the header and footer.

3. Specify the header:

    a. On a new line, type **\fancyhead[*location*]{*text*}** where *location* specifies the page and field of the header and *text* specifies the information you want to appear.
    For example, the command \fancyhead[LE,RO]{\thepage} places the page number in the left field of the header on even (left) pages and in the right field of the header on odd (right) pages.
    b. Repeat step a as needed to specify the header fields on both even and odd pages.

4. Specify the footer:

    a. On a new line, type **\fancyfoot[*location*]{*text*}** where *location* specifies the page and field of the footer and *text* specifies the information you want to appear.
    b. Repeat step a as needed to specify the footer fields on both even and odd pages.

5. Choose **OK**.

Additional instructions about using the *fancyhdr* package appear earlier in this manual. See page 7 for instructions about adding a rule under a header, page 11 for instructions about moving page numbers, and page 8 for instructions about specifying header and footer information. The package documentation contains complete instructions and useful diagrams of LaTeX page layout elements. The installation program places this package in the TCITeX/TeX/LaTeX/contrib/supported/fancyhdr directory.

Fix2col

The *fix2col* package modifies the LaTeX two-column output routine in two specific ways. First, it improves the handling of marks. If the first column of a two-column page contains marks, the package instructs LaTeX to take \firstmark from the first column rather than discarding it. Second, the package improves the handling of floating objects by anchoring both one- and two-column floating objects in a single sequence. Without the package, LaTeX may anchor two-column floating objects after those with a single column.

No action is required beyond adding the package to your two-column document. No package options are available for the *fix2col* package, which is installed in the TCITeX/TeX/LaTeX/contrib/supported/carlisle directory.

Flafter

Ordinarily, LaTeX tries to place floating objects at the top of the page, regardless of where the reference to the object occurs. The *flafter* package overrides the ordinary placement to force LaTeX to print the floating object after the reference to it.

No action is required beyond adding the *flafter* package to your document. No package options are available. The package is installed in the `TCITeX/TeX/LaTeX/base` directory as part of the Standard LaTeX Base System.

Float

Instead of being anchored at a specific point in a LaTeX document, floating objects are movable. LaTeX determines their best placement and final position in the typeset document after taking into consideration line and page breaks and division headers. You can suggest a preferred placement to LaTeX by specifying t (top of a page), b (bottom of a page), p (on a page of floating objects), or h (where the floating object appears in the document file, if possible).

The *float* package improves the interface for defining and placing floating objects by defining the H (Here) placement option of the superseded Here package. The H option tells LaTeX to place a floating object where it appears in the document file, even if the placement is typographically ill advised. If not enough room remains on the page to hold the object, LaTeX moves it to the next page, leaving blank space on the page before.

SWP and *SW* work differently with the *float* package depending on how you save your document. If you save it as an *SWP* or *SW* document, the program automatically uses the H placement option when you select only the Here placement for a floating graphic. If you save the document as a Portable LaTeX document, the program ignores the H option. Using the package reduces the likelihood that LaTeX will accumulate too many floating objects to typeset your document correctly.

▶ **To force the placement of floating objects where they appear in the document file**

1. Add the *float* package to your document.

2. Enter a floating object. (Use the fragment named Table - (4x3, floating) to enter a table that floats.)

3. Edit the properties of the floating object to specify only the Here placement:
   - If the object is a graphic,
     i Edit the properties of the graphic to open the **Graphic Properties** dialog.
     ii Choose the **Layout** tab.
     iii In the **Placement** area, select **Floating**, check **Here**, uncheck the other three placement options, and then choose **OK**.

     or
   - If the object is a table,
     i Edit the [B] box in the Table - (4x3, floating) fragment, which contains the command \begin{table}[tbp] \centering.
     ii Change the [tbp] entry to **[H]** and choose **OK**.

The *float* package implements two float styles: boxed and ruled floats. Boxed floats are surrounded by a box extending from the right to the left margin, regardless of the size of the floating object. Ruled floats are introduced by a horizontal line; have another line under the caption, if any; and are followed by a third line, all extending from the right to the left margin. You must define the float style and the floating object in an encapsulated TeX field. The package also provides a way to define your own types of floating objects, which can be ruled or boxed. See the documentation for the float package for more information.

The *float* package has no options. The installation program places the package in the `TCITeX/TeX/LaTeX/contrib/supported/float` directory.

Fontsmpl

The *fontsmpl* package produces a test of a font family, such as Computer Modern or Times, showing the font in use in sample text, a table of contents, and a sample of commands. You can use the package to print a sample of the font currently in use in your document or you can open and typeset a sample document that produces a sample of the font family you indicate.

▶ **To print a sample of the font currently in use**

1. Add the *fontsmpl* package to your document.

2. Place the insertion point where you want to sample the font.

3. Enter an encapsulated TeX field.

4. Type **fontsample** and choose **OK**.

5. Typeset preview the document.

▶ **To produce a font sample with the sample document**

1. Open the file `fontsmpl.tex` in the `TCITeX/TeX/LaTeX/required/tools` directory.

2. Typeset preview the document.

3. When LaTeX prompts you for a font family, type the font family name, such as **cmr** for Computer Modern or **times** for Times Roman.

4. Press ENTER.

5. Scroll through the document displayed in the TrueTeX Previewer to examine the appearance of the font.

The package has no options. The installation program places the package in the `TCITeX/TeX/LaTeX/required/tools` directory.

Footmisc

The *footmisc* package was designed for customizing footnotes. The package options address the format, positioning, and numbering of single and multiple footnotes. The package provides for using symbols instead of footnote numbers. It also provides some debugging tools.

▶ To customize footnotes

1. Add the *footmisc* package to your document.

2. Modify the package options to select the option you want.

3. Choose OK.

The package is installed in the `TCITeX/TeX/LaTeX/contrib/supported/footmisc` directory.

Ftnright

The *ftnright* package formats footnotes for two-column documents. It prints all footnotes that occur on a page at the foot of the right-hand column. Footnotes appear in smaller type than that used for the text. The footnote numbers are set on the baseline rather than as superscripts. The text and the footnotes aren't separated by a line. When you add the *ftnright* package to your document, specify a two-column layout with the document class options instead of the *multicol* package (see page 121). The *ftnright* package doesn't work successfully with *multicol*.

After adding the package to your document, no further action is required. The package has no options. The *ftnright* package is installed in the `TCITeX/TeX/LaTeX/required/tools` directory as part of the Standard LaTeX Tools Bundle.

Geometry

The *geometry* package provides a simple way to customize the page layout of your document. If the shell you're using provides a largely adequate layout, you may be able to use the *geometry* package to customize the shell so that it meets all your typesetting requirements. In particular, the easiest way to change the typeset margins for a document is to add the *geometry* package and then include in the document preamble a command to change the margins.

You can use the package to specify portrait or landscape orientation, margins, margin offsets for two-sided printing, elimination of the space for headers and/or footers, paper size, horizontal and vertical offsets, and many other typesetting details. The package uses automatic completion of layout dimensions; if you don't specify all dimensions, the package supplies the remainder automatically. The *geometry* package also uses auto-centering and auto-balancing mechanisms so that you can use simple, minimal descriptions to define the page layout you want. For example, you can set all the margins to 3 centimeters without any space for headers or footers using this command in the preamble of your document:

\usepackage[margin=3cm,noheadfoot]{geometry}

The package uses the "key val" (key = value) interface. As seen in the above example, command arguments consist of key val options, separated by commas. Options are usually order-dependent. You can specify commands in multiple lines.

With the **Options and Packages** command on the **Typeset** menu, you can set options for the *geometry* package. The options affect page orientation, paper size, print side, margin notes, headers and footers, and magnification. You can make additional specifications with TeX commands in the preamble of your document. See Chapter 1 "Tailoring Typesetting to Your Needs" to learn how to use the *geometry* package to change margins, page orientation, header and footer space, and paper size. The package documentation contains additional instructions and a good illustration of LaTeX page layout concepts. The package is installed in the `TCITeX/TeX/LaTeX/contrib/supported/geometry` directory.

Graphicx

The package is one of three packages—with *color* (see page 97) and *graphics*—included as part of the Standard LaTeX Graphics Bundle. Although identical to the *graphics* package in function, the *graphicx* package has an interface that is easier to use and more powerful.

The *graphicx* package implements LaTeX support for including graphics files, rotating parts of a page, and scaling parts of a page. The package depends on having a typeset output driver that can produce these effects. The TrueTeX Previewer provided with *SWP* and *SW* doesn't support rotation; use a different DVI previewer and print driver if you want to use the graphicx package to rotate parts of a page in a DVI file. However, PDF viewers support rotation, so you can use the package to rotate parts of a page in typeset PDF files.

You provide additional information to the package with the \includegraphics commands placed in TeX fields in the body of your document. The command has optional arguments that define the type of graphic; its desired size, shape, and angle of rotation; and the size and shape of the box surrounding the graphic. See page 40 for instructions about using the *graphicx* package to correct problems with typesetting PostScript graphics. The package is installed in the `TCITeX/TeX/LaTeX/required/graphics` directory.

Harvard

The *harvard* package is a family of seven BibTeX bibliography styles:

Bibliography Style	Meets format requirements of
AGSM	*Australian Government Style Manual*
APSR	*American Political Science Review*
DCU	*Design Computing Unit, Department of Architectural and Design Science, University of Sydney*
JMR	*Journal of Management Research*
J Physics B	*Journal of Physics B*
Kluwer	*Kluwer Academic Publishers*
Nederlands	Dutch conventions

Although the format differs with each style, all seven styles implement standard parenthetical citations:

> The definitive work on the subject (Medd 2002)....

Citations can be complete or incomplete. Complete citations can be used as nouns:

> Medd (2002) proves that....

or as possessives:

> Medd's (2002) definitive work on the subject....

Incomplete citations can contain the author's name without the date:

> The definitive work on the subject (Medd)....

or the date without the name:

> The definitive work on the subject (2002)....

The options available through the **Options and Packages** command on the **Typeset** menu determine the type of citation (full, abbreviated, or full for the first citation and abbreviated thereafter), type of parentheses, and style of the citations and bibliography. Although the package was designed for use with BIBTeX bibliographies, you can also use it with a manual bibliography. The `PackageSample-harvard-manual.tex` file in the `SWSamples` directory illustrates the use of the package with a manual bibliography. The Standard LaTeX Article (Harvard) shell contains more information.

▶ **To use a harvard bibliography style with a BIBTeX bibliography**

1. Add the *harvard* package to your document and select any options you want.

2. From the **Typeset** menu, choose **Bibliography Choice**.

3. Check **BibTeX** and choose **OK**.

4. Insert citations as needed in your document.

5. Specify the harvard .bst file you want:
   a. Place the insertion point where you want the bibliography to appear.
   b. From the **Insert** menu, choose **Typeset Object** and then choose **Bibliography**.
   c. Select the BIBTeX database file you want to use.
   d. Scroll down the **Style** list to select one of the harvard .bst files and choose **OK**.

6. Save the document.

7. On the Typeset toolbar, click the Typeset DVI Compile button or, from the **Typeset** menu, choose **Compile**.

8. Check **Generate a Bibliography** and choose **OK**.

9. Typeset preview your document.

The *harvard* package is intended to be used with BIBTEX bibliographies. You can also use it with manually created bibliographies by inserting \harvarditem commands in TEX fields to create the bibliography list.

▶ **To use a harvard bibliography style with a manual bibliography**

1. Add the *harvard* package to your document and select any options you want.

2. From the Typeset menu, choose Bibliography Choice.

3. Check Manual and choose OK.

4. Create citations as needed in your document.

5. Begin the bibliography list:

    a. Place the insertion point where you want the bibliography to appear in your document.
    b. Enter an encapsulated TEX field.
    c. In the entry area, type **begin{thebibliography}**{*x*} where *x* is the longest label in the bibliography list and determines the indention of the list items.
    d. Choose OK.

6. For each item in the list:

    a. Enter an encapsulated TEX field.
    b. Name the field with the key to provide a visual reminder of the key.
    c. In the entry area, type **harvarditem**{*citation*}{*year*}{*key*} where *citation* is the information to be cited in the text except for the year, *year* is the year to be cited, and *key* is the key of the reference used in the citation.
    d. Choose OK.
    e. Type the reference in full as you want it to appear in the list of references.

7. End the bibliography list:

    a. Place the insertion point after the last item.
    b. Enter an encapsulated TEX field.
    c. In the entry area, type **end{thebibliography}** and choose OK.

8. Save and typeset compile the document.

See an example of the package in use in the `Harvard.tex` file in the `TCITeX/doc/latex/contrib/supported/harvard` directory of your program installation. The package is in the `TCITeX/TeX/LaTeX/contrib/supported/harvard` directory.

Hhline

The *hhline* package works with the *array* package (see page 86) to implement fine control of single and double horizontal lines (like \hline and \hline\hline) within typeset tables, as shown here:

a	b	c	d
s	w	p	5.0
1	2	3	4
5	6	7	8

Note that you can achieve similar results in SWP and SW without the *hhline* package:

a	b	c	d
s	w	p	5.0
1	2	3	4
5	6	7	8

However, when you use double lines around the outside of a table, the *hhline* package may produce a more pleasing appearance at the outside corners.

The lines are governed by the package command \hhline. The command arguments are tokens, or symbols, that indicate the absence, presence, and width of a horizontal line and whether or not it breaks or is broken by an intersecting vertical line. See the package documentation for a discussion of the tokens.

▶ **To create horizontal lines within a tabular environment**

1. Add the *hhline* package to your document.

2. Enter an encapsulated TeX field.

3. In the entry area, type **begin{tabular}** to define a tabular environment.

4. Enter the tabular content.

5. Between rows of content, use \hhline commands with appropriate tokens to define the lines you want.

6. Type **end{tabular}** to end the environment and choose **OK**.

No package options are available. The *hhline* package is installed in the `TCITeX/TeX/LaTeX/required/tools` directory as part of the Standard LaTeX Tools Bundle.

Hyperref

From LaTeX cross-referencing commands, including the table of contents, bibliographies, and page-references, the *hyperref* package creates \special commands that a driver can turn into hypertext links. The package also provides commands with which you can write ad hoc hypertext links, including links to external documents and URLs. Note that the TrueTeX Previewer provided with SWP and SW doesn't support the package, but PDF viewers do, so you can use *hyperref* to maintain active LaTeX cross-references in PDF files that you typeset from your document. When you add the *hyperref* package to your document, move it to the bottom of the **Packages in Use** list.

The package has macros and options available; see the extensive package documentation for information about the commands. The *hyperref* package is installed in the `TCITeX/TeX/LaTeX/contrib/supported` directory.

Hyphenat

The *hyphenat* package disables hyphenation in parts of your document or in the document as a whole. Also, it enables hyphenation of words containing nonalphabetic characters, such as underscores, and text that is set in monospaced fonts.

The two available package options turn hyphenation on and off and control hyphenation for monospaced fonts. Note that when you select the **None** option, which prevents all hyphenation, you may get LaTeX messages about bad line breaks and overfull boxes when you compile your document. For more information, see page 31. The *hyphenat* package is installed in the `TCITeX/TeX/LaTeX/contrib/supported/hyphenat` directory.

Indentfirst

The *indentfirst* package indents the first line of all sections by the usual paragraph indention. Other than adding the package, no action is required. No package options are available. The package works successfully with all standard document classes. It is installed in the `TCITeX/TeX/LaTeX/required/tools` directory as part of the Standard LaTeX Tools Bundle.

Inputenc

The *inputenc* package maps certain characters to their corresponding TeX macros according to the encoding option you select. Available options include ISO Latin-1, ISO Latin-2, and others.

The package has no effect in *SWP* or *SW*, but when a document containing the package is opened in a text editor, encoded characters appear correctly rather than as TeX code. For example, when the ISO Latin-1 option is selected as the font encoding scheme, the character Ã appears in a text editor as Ã instead of as `\~{A}`. The *inputenc* package may simplify collaboration on documents written in some non-English languages.

See the package documentation for additional information. The package is installed in the TCITeX/TeX/LaTeX/base directory.

Latexsym

The *latexsym* package provides 11 mathematical symbols that were originally defined in LaTeX 2.09 but are no longer defined in the New Font Selection Scheme. The symbols are

$$\Cup \bowtie \Box \Diamond \leadsto \sqsubset \sqsupset \lhd \unlhd \rhd \unrhd$$

These symbols are also provided by the *amsfonts* and *amssymb* packages (see page 83). Because *SWP* and *SW* call the *amsfonts* package automatically, ordinarily you don't need to add the *latexsym* package to your document to obtain the symbols.

No package options are available. The *latexsym* package is installed in the `TCITeX/TeX/LaTeX/base` directory.

Layout

The *layout* package illustrates the LaTeX layout of the current document with a figure similar to the one on page 117. The circled callouts refer to the accompanying table of standard LaTeX page layout values including paperwidth, topmargin, oddsidemargin, textheight, textwidth, headheight, and others. The package is useful for refining the layout of your document.

The options available for the package through the **Options and Packages** command on the **Typeset** menu determine the language in which the layout is printed and provide debugging aids.

▶ **To draw the layout of a document**

1. Add the *layout* package to your document.
2. Place the insertion point anywhere in the body of your document.
3. Enter a TeX field.
4. Type **\layout** and choose **OK**.

   LaTeX draws the layout in your document immediately after the command.

   The *layout* package doesn't work for documents created with Style Editor shells. It is installed in the `TCITeX/TeX/LaTeX/required/tools` directory as part of the Standard LaTeX Tools Bundle.

Lineno

The *lineno* package provides line numbers on typeset paragraphs. By default, the numbers run continuously through the document, but you can reset the line numbers on each page. The package provides a way to create a reference to a particular line number using the cross-reference mechanism.

▶ **To add line numbers to a document**

1. Add the *lineno* package to your document.
2. Modify the package options to select the options you want and choose **OK**.
3. From the **Typeset** menu, choose **Preamble** and click the mouse in the entry area.
4. On a new line, type **\linenumbers** and choose **OK**.

The package is installed in the `TCITeX/TeX/LaTeX/contrib/supported/lineno` directory.

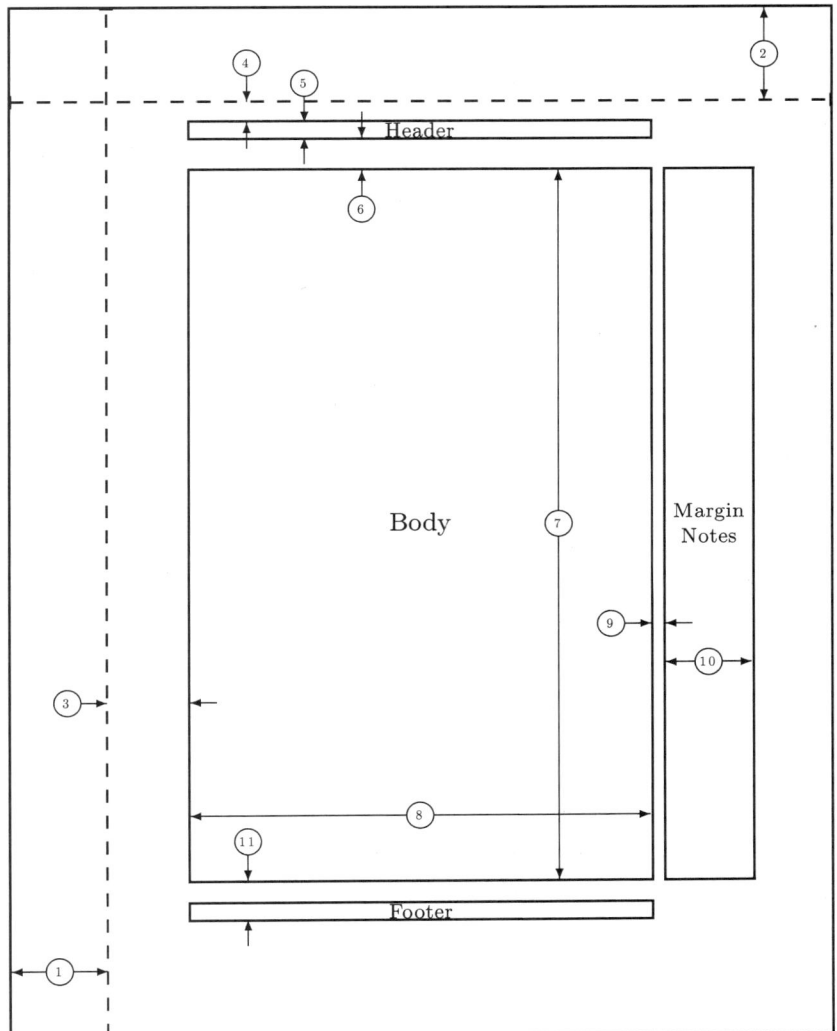

1	one inch + \hoffset	2	one inch + \voffset
3	\oddsidemargin = 62pt	4	\topmargin = 16pt
5	\headheight = 12pt	6	\headsep = 25pt
7	\textheight = 550pt	8	\textwidth = 345pt
9	\marginparsep = 11pt	10	\marginparwidth = 65pt
11	\footskip = 30pt		\marginparpush = 5pt (not shown)
	\hoffset = 0pt		\voffset = 0pt
	\paperwidth = 614pt		\paperheight = 794pt

Longtable

The package defines a longtable environment that is a multipage version of tabular. The tables produced by the package can be broken between, but not within, rows by standard TEX page-breaking algorithms. The package maintains consistent column widths from page to page, unlike the *supertabular* package (see page 139). The package provides customized captions on the first and subsequent pages of the table, but you may find the *caption2* package more convenient to use for that purpose (see page 92).

The options available through the **Options and Packages** command on the **Typeset** menu aid in debugging. The package commands define and customize the longtable environment. Commands are available to split the table into a series of *chunks* for easier management by TEX. Additional commands position the table, define rows, define breaks, and make footnotes available within a table environment. The *longtable* package doesn't require the *array* package, but if *array* is added to the document, the *longtable* package uses extended features. See the package documentation for a discussion of how to use each option, command, and parameter.

To use the *longtable* package in your document, you must define the entire longtable environment in an encapsulated TEX field. Also, you may have to process the document through LaTeX three or four times to achieve the correct appearance of columns in a longtable environment. Using the package requires a solid knowledge of LaTeX. For more information, see the package documentation and the `PackageSample-longtable.tex` file in the `SWSamples` directory of your program installation. The *longtable* package is installed in the `TCITeX/TeX/LaTeX/required/tools` directory as part of the Standard LaTeX Tools Bundle.

Lscape

The *lscape* package creates a landscape environment within which LaTeX rotates text 90 degrees. Note that PDF viewers support rotation, so you can use the lscape package to create rotated text in typeset PDF files. The TrueTEX Previewer provided with *SWP* and *SW* doesn't support rotation; you must use a different DVI previewer and print driver if you want to use the lscape package to rotate text in a DVI file. Although you can use a different driver, we recommend that you leave the driver defaults unchanged. The local LaTeX installation sets the driver defaults. If you leave the configuration unchanged for your document, you can compile it without changes in another LaTeX environment.

You can define the landscape environment with commands inserted in TEX fields. The *portland* package produces similar results (see page 126).

▶ **To define a landscape environment**

1. Add the *lscape* package to your document.

2. Place the insertion point where you want landscape orientation to begin in your document.

   Place the commands carefully so that LaTeX rotates the current page before adding the page header and footer.

3. Enter an encapsulated TEX field.

4. Type \begin{landscape}.

5. Choose OK.

6. Place the insertion point where you want to return to portrait orientation.

7. Enter an encapsulated TeX field.

8. Type \end{landscape}.

9. Choose OK.

Many driver options are available for *lscape* through the Options and Packages command on the Typeset menu. The package is installed in the `TCITeX/TeX/LaTeX/required/graphics` directory as part of the Standard LaTeX Graphics Bundle.

Ltxtable

The *ltxtable* package is a combination of two other packages—*longtable* and *tabularx*—that are part of the Standard LaTeX Tools Bundle. The package produces multipage tables with column widths calculated automatically to meet a total specified table width.

Using the package involves placing the longtable environment in a file all its own, then including it in the main document. The command syntax for inputting the file is

$$\textbf{\textbackslash Ltxtable}\{\textit{width}\}\{\textit{file}\}$$

No options are available for *ltxtable*. The package is installed in the `TCITeX/TeX/LaTeX/contrib/supported/carlisle` directory.

Makeidx

The *makeidx* package creates an index for your document based on information in \index commands in the text. *SWP* and *SW* automatically add the *makeidx* package to your document when you create the index, but you must add the package yourself if you plan to save your document as a Portable LaTeX file.

▶ **To create an index with makeidx**

1. Create index entries throughout your document, as needed.

2. Place the insertion point where you want the index to appear.

3. Import the index fragment.

4. Typeset compile your document and choose Generate an Index or, from the Typeset menu, choose Tools and then choose Run MakeIndex.

The program adds the *makeidx* package to your document and the \makeindex command to the document preamble, and generates the index.

Note that if you plan to save your document as a Portable LaTeX file, you must add the *makeidx* package to your document.

No package options are available. The package is installed in the `TCITeX/TeX/LaTeX/base` directory.

Mathpple

See PSNFSS Packages on page 127.

Mathptm

See PSNFSS Packages on page 127.

Mathptmx

See PSNFSS Packages on page 127.

Mathtime

The package enables the use of PostScript New Font Selection Scheme (PSNFSS) MathTime fonts, including MathTime Plus fonts, in LaTeX documents. When you typeset your document, MathTime fonts provide ligatures and improved kerning in Times text. The *mathtime* package implements mathematics set in Times and calls the *times* package directly, so that documents containing both mathematics and text can use Times throughout, rather than a combination of font families. The result is a more portable document with a nicer appearance. See also the *times* package, page 128. When you add the *mathtime* package to your document, move it near the bottom of the **Packages in Use** list.

The *mathtime* package has several options for font use available through the **Options and Packages** command on the **Typeset** menu. The **No TS1** option must be set for use with the TrueTeX Formatter included with *SWP* and *SW*.

▶ **To use MathTime fonts in a document**

1. On the Typeset toolbar, click the Options and Packages button ▣ or, from the **Typeset** menu, choose **Options and Packages** and then choose the **Package Options** tab.

2. Select the font package currently in use.

3. Choose **Remove**.

4. Add the *mathtime* package to your document.

5. Modify the package to select the **No TS1** option.

   The package won't work correctly without this setting.

6. Choose **OK**.

The package is installed in the `TCITeX/TeX/LaTeX/required/mathtime` directory. See the online Help and the package documentation for more information

Multicol

about the implementation of the mathtime package in a TrueTeX environment. See also the file `OptionsPackagesLaTeX.tex` in the `SWSamples` directory.

The *multicol* package implements up to 10 columns of text in the multicols environment and balances the length of the final columns for a nice appearance. The package permits both single- and multicolumn formats on the same page. It places footnotes across the bottom of the page rather than under each column. (Thus, it is incompatible with the *ftnright* package, page 110.) Note that within the multicolumn environment, only page-wide floating elements are permitted and not those in a selected column.

▶ **To create multiple columns of text**

1. Add the *multicol* package to your document.
2. Place the insertion point where you want multiple columns to begin.
3. Enter an encapsulated TeX field.
4. Type **begin{multicols}**{*x*} where *x* is the number of columns you want.
5. Choose **OK**.
6. Place the insertion point where you want multiple columns to end.
7. Enter an encapsulated TeX field.
8. Type **end{multicols}** and choose **OK**.

The package has a debugging option available through the **Options and Packages** command on the **Typeset** menu. The use of the *multicol* package can occasionally cause LaTeX errors related to line spacing and footnote numbering in *SWP* and *SW* documents. The package is installed in the `TCITeX/TeX/LaTeX/required/tools` directory as part of the Standard LaTeX Tools Bundle.

Newapa

This package is required when you use the BibTeX bibliography style `newapa.bst`. Other than adding the package to your document, no action is required. No options are available for the package, which is installed in the `TCITeX/TeX/LaTeX/contrib/other/bibtex` directory.

Newpnts

See Points and Newpnts on page 125.

Nomencl

The *nomencl* package produces a nomenclature list or list of symbols for your document by using TeX instructions inserted throughout the document as input for the makeindx program.

Generating a nomenclature list is a multistep process involving adding the package to your document, identifying each symbol to be included in the list, indicating where the list should appear in the document, and finally running makeindx and compiling your document with LaTeX.

▶ **To create a nomenclature list**

1. Add the *nomencl* package to your document.

2. If your document meets these three conditions:
   - you are using the SWP or SW output filter (not the Portable LaTeX filter)
   - the highest division level in your document is section
   - the chapter division isn't used

   then

   a. Save, close, and reopen the document.
   b. From the **Typeset** menu, choose **Preamble**.
   c. Click the mouse in the entry area and scroll to the bottom.
   d. After the line \input{tcilatex}, add a new line and type **\let\chapter\undefined**.
   e. Choose **OK**.

3. Add the \makeglossary command to the document preamble.

   a. From the **Typeset** menu, choose **Preamble** and click the mouse in the entry area.
   b. Scroll to the end of the preamble commands and begin a new line.
   c. Type **\makeglossary** and choose **OK**.

4. Add \nomenclature commands in the text for each symbol to be included in the nomenclature list:

   a. Place the insertion point immediately after the first use of the symbol.
   b. Enter an encapsulated TeX field.
   c. In the entry area, type **\nomenclature{x}{y}** where x is the symbol you want to appear in the list and y is the corresponding definition.

5. Add the \printglossary command.

   a. Place the insertion point where you want the nomenclature list to appear in your document.
   b. Enter a TeX field.
   c. In the box, type **\printglossary** and choose **OK**.

6. Save and compile your document.

   The compilation yields a .glo file with the same name as your document and places it in the same directory.

7. Run makeidx.

   The makeidx package uses the .glo file as an input file. It creates an output file with a .gls extension and the same name and directory as your document. The .gls file contains the correctly ordered nomenclature list.

a. From the Windows **Start** menu, choose **Run**.

b. In the **Open** box, type this command (where line breaks occur in this instruction, enter a space):
**c:\swp50\TCITeX\TrueTeX\obsolete\makeindex -o
c:\swp50\docs*filename*.gls -s
c:\swp50\TCITeX\TeX\LaTeX\contrib\supported\nomencl\nomencl.ist
c:\swp50\docs*filename*.glo**
Substitute the correct file name for the `.glo` and `.gls` files. These instructions reflect a Version 5 *SWP* installation. Correct the path names for your installation, if necessary.

c. Choose **OK**.

8. Typeset compile the document file from outside *SWP* or *SW*.

   Note that if you compile using *SWP* or *SW*, the compiler won't find the `.gls` file and won't include the nomenclature list in the typeset document.

   a. From the *SWP* or *SW* program group, choose the TrueTeX Formatter.
   b. Select the file and choose **OK**.
      If your document contains a table of contents or cross-references, you may need to compile it two or three times.

The available package options include language choices and the use of equation and page references. See more information on page 21 and additional documentation for the *nomencl* package in the `PackageSample-nomencl.tex` file in the `SWSamples` directory of your program installation. The package is in the `TCITeX/TeX/LaTeX/contrib/supported/nomencl` directory.

Nopageno

The *nopageno* package provides a simple way to remove page numbers from both the opening pages and the normal pages of all classes of LaTeX documents. Other than adding the package to your document, no action is necessary. No package options are available. The package installs in the `TCITeX/TeX/LaTeX/contrib/supported/carlisle` directory.

Numinsec

The *numinsec* package provides numbering of equations, figures, and tables within sections. Part of the SIAM document class distribution, the package is useful for other types of documents. The package adds the section number to the numbers of equations, figures, and tables. Thus, if the first equation in section 2 is the eighth in the document, it carries the number 2.1 instead of 8.

No action is required other than adding the package to your document. The package has no options. The *numinsec* package is installed in the `TCITeX/TeX/LaTeX/contrib/other/siam` directory.

Overcite

The *overcite* package produces and superscripts compressed, sorted lists of numerical citations occurring in the text. The package prints the citation numbers in ascending order, separated by a comma and a space; three or more consecutive numbers appear as a range. Thus, a citation of [7,3,5,9] becomes [3,5,7,9] and a citation of [3, 6, 5, 4] becomes [3−6].

▶ **To order and compress citations created with SWP or SW citations**

1. Add the *overcite* package to your document.

2. Place the insertion point where you want the citation to occur.

3. Choose [icon] or, from the **Insert** menu, choose **Typeset Object** and then choose **Citation**.

4. In the **Key** box, enter the key for the reference you want to cite or select the key from the dropdown list.

5. Select the contents of the **Key** box and copy the selection to the clipboard.

6. In the **Key** box, overwrite the selection with the key for the next reference.

7. Type a comma and press the spacebar.

8. Paste the contents of the clipboard to the **Key** box.

9. Repeat steps 5–8 until you have entered all the references you need, and then choose OK.

When you typeset your document, LaTeX orders the list. Whenever the list includes three references in numerical sequence, LaTeX compresses them. Command variations are available; see the package documentation for more information about using the package. In addition, the options available for *overcite* provide spacing and sorting refinements for citations.

The *overcite* package is one of four compatible packages—including *cite* (see page 96), *drftcite* (see page 101), and *chapterbib* (see page 93)—installed in the `TCITeX/TeX/LaTeX/contrib/supported/cite` directory.

Paralist

The *paralist* package provides new list environments for itemized, description, and enumerated lists. With the package, lists can be typeset within paragraphs, as paragraphs in themselves, and in a compressed format. The package allows adjustment of the space between list items in the compressed format. The package also provides arguments for formatting labels in most of the list environments. The package incudes a configuration (`.cfg`) file that makes standard list environments typeset as if they were the compressed list environments defined by the package. Although the .cfg file isn't part of the default package, the package allows adding a .cfg file. The package may conflict with the *babel* package.

The package is installed in the `TCITeX/TeX/LaTeX/contrib/supported/paralist` directory.

Pifont

See PSNFSS Packages on page 127.

Points and Newpnts

The *points* and *newpnts* packages were developed to help typeset exams. Both packages place point values for exam questions in the left margin. The *newpnts* package places the values slightly farther into the margin for less confusion with list item numbers. The packages also provide a way to resume list numbering that has been interrupted by text. The two features can be used independently.

No options are available for the packages, but you can enter the necessary macros in TeX fields. The packages are distributed with *SWP* and *SW* but, unlike the other packages included with the program, are not available on CTAN. The packages are installed in the `TCITeX/TeX/LaTeX/SWmisc` directory.

▶ **To place a point value for an exam question in the margin**

1. Add the *points* or the *newpnts* package to your document.

2. Create the list of exam questions.

   Note that you don't have to use a numbered list.

3. For each question,

   a. Place the insertion point at the beginning of the list item, immediately to the right of the lead-in object.
   b. Enter an encapsulated TeX field.
   c. Type **POINTS**{*x*} where **x** is the point value for the question.
   d. Choose OK.

▶ **To resume numbering of an interrupted numbered list**

1. Add the *points* or the *newpnts* package to your document.

2. Place the insertion point at the end of the paragraph immediately after which you want the list to begin.

   The paragraph can otherwise be empty.

3. Enter an encapsulated TeX field.

4. In the entry area, type **setcounter**{**enumi**}{*x*}**RESUME** where *x* is a value one less than the starting number for the list.

If you want your list to start with 1, use \setcounter{enumi}{0}\RESUME. Note that the command is case-sensitive.

5. Choose **OK**.

6. Create a numbered list.

7. Interrupt the list with an unnumbered paragraph.

8. Place the insertion point at the end of the unnumbered paragraph.

9. Enter an encapsulated TeX field.

10. In the entry area, type **\RESUME** and choose **OK**.

11. Enter the remaining numbered list items.

    LaTeX resumes the list with the next sequential list item number.

Portland

The *portland* package implements changing from portrait to landscape orientation and back within your *SWP* or *SW* document. No special drivers are required, but you may need to change the orientation settings for your printer so that your document prints properly. If you have a single page with an orientation different from that of the rest of the document, you may need to print it separately after changing the printer settings accordingly.

You can define portrait and landscape environments with simple commands inserted in TeX fields. No package options are available.

▶ **To define a landscape or portrait environment**

1. Add the *portland* package to your document.

2. Place the insertion point where you want the orientation to change in your document.

3. Enter a TeX field.

4. Type **\landscape** to change the page layout orientation from portrait to landscape.

   or

   Type **\portrait** to change the page layout orientation from landscape to portrait.

5. Choose **OK**.

The *portland* package is installed in the `TCITeX/TeX/latex209/contrib/misc` directory of your Version 3.5 or later installation. The `PackageSample-portland.tex` file in the `SWSamples` directory of your Version 4.0 or later program installation contains additional information. If you have an earlier version you can download both the package and the sample file from our website. See also the *lscape* package on page 118.

PSNFSS Packages

The PostScript NFSS bundle is a required part of LaTeX. The bundle provides a complete working setup of the LaTeX font selection scheme for use with PostScript fonts. The packages are installed in the `TCITeX/TeX/LaTeX/required/psnfss` directory.

The font packages included in the bundle completely replace one or more of the font families used by LaTeX for roman, sans serif, typewriter, or mathematics fonts. If the PostScript fonts are present on your system, adding the *PSNFSS* packages changes the default font families as shown in the table below. The top row indicates the default (Computer Modern) font family. A blank indicates that the package doesn't change the corresponding default font family.

	Font Family			
Package	**roman**	**sans serif**	**typewriter**	**math**
(none)	CM Roman	CM Sans Serif	CM Typewriter	≈ CM Roman
Mathptmx	Times			≈ Times
mathptm	Times			≈ Times
mathpple	Palatino			≈ Palatino
helvet		Helvetica		
avant		Avant Garde		
courier			Courier	
chancery	Zapf Chancery			
times	Times	Helvetica	Courier	
palatino	palatino	Helvetica	Courier	
bookman	Bookman	Avant Garde	Courier	
newcent	New Century Schoolbook	Avant Garde	Courier	
utopia	Utopia			
charter	Charter			

Other than installing the packages, no change is required to make the default replacements noted in the table.

Helvet

If you have the Helvetica PostScript font installed on your system, adding the *helvet* package changes the default sans serif font family to Helvetica. (Note that Helvetica is slightly larger that other typefaces. The package documentation explains how to adjust for the difference.) If the font isn't installed on your system, adding the package changes the default sans serif font to the Windows Arial font. The other font packages—*avant, courier, chancery, times, palatino, bookman, newcent, utopia, charter*—work the same way

Adding the package doesn't affect the default serif font. If you want your entire document to print in the Helvetica font, modify the preamble of your document.

▶ **To change the default font**

1. From the **Typeset** menu, choose **Preamble**.

2. Click the mouse in the entry area and start a new line after the last entry.

3. Type **\renewcommand{\familydefault}{\sfdefault}** and choose **OK**.

The other font packages—*avant, courier, chancery, times, palatino, bookman, newcent, utopia, charter*—work essentially the same way

Mathpple
The package changes the default roman font family to Adobe Palatino. The virtual mathpple fonts are used for mathematics. Package options include slanted uppercase Greek letters, upright Δ and Ω, italic bold mathematical symbols, and large mathematical symbols. Using the *mathpple* package with AMS symbols requires special attention; please refer to the package documentation.

Mathptm
The *mathptm* package, an earlier version of the *mathptmx* package, sets lowercase Greek upright. The package requires the cmex9 font, which is not always available. The *mathptmx* package is preferred.

Mathptmx
The *mathptmx* package changes the default roman font family to Adobe Times. The virtual mathptmx fonts are used for mathematics. The package scales large mathematical symbols to fit the base font size. Bold math fonts are not supported.

Pifont
The *pifont* package supports symbol fonts and provides commands for using the Zapf Dingbats font; see the package documentation for details. The package also provides an interface to other font families.

Times
The *times* package implements the use of the NFSS Times font for text but leaves mathematics in the Computer Modern fonts. The package produces ligatures and improves kerning, and its use creates more portable document files. See also the *mathtime* package on page 120. This example illustrates the mathematics and text produced by the *times* package:

By the triangle inequality for integrals and the above inequalities, for $n \geq N$;

$$\left| \int_C \left[f(z) - \sum_{k=0}^{n} a_k z^k \right] dz \right| \leq \epsilon \cdot (\text{length of } C)$$

Since ϵ is arbitrary, the limit is zero.

▶ **To use the times package**

1. On the Typeset toolbar, click the Options and Packages button or, from the Typeset menu, choose Options and Packages and then choose the Package Options tab.

2. Select the font package currently in use and choose Remove.

3. Add the *times* package to your document and choose OK.

No options are available. See the online Help and the package documentation for more information about the implementation of the *times* package in a TrueTeX environment.

Remreset

The *remreset* package removes a counter from the reset list controlled by a second counter. In other words, the package prevents a designated counter from being reset when LaTeX increments a second counter. For example, most reports normally reset the equation number at the beginning of each chapter. You can use this package to remove the reset so that equations are numbered sequentially throughout the report.

▶ **To prevent a counter from being reset**

1. Add the *remreset* package to your document.

2. From the Typeset menu, choose Preamble and click the mouse in the entry area.

3. Add the commands

   \makeatletter
   \@removefromreset{*x*}{*y*}
   \makeatother

   where *x* is the counter you don't want to reset, such as footnote or theorem, and *y* is the controlling counter, such as chapter or section.

4. Choose OK.

The package command \@removefromreset is the opposite of the standard LaTeX command \@addtoreset.

No package options are available. The package is installed in the `TCITeX/TeX/LaTeX/contrib/supported/carlisle` directory.

Revsymb

The *revsymb* package defines the lambdabar symbol λ and other symbols unique to REVTeX 4. Other than adding the package to your document, no action is required. The *revsymb* package is part of the REVTeX 4 distribution and is installed in the `TCITeX/TeX/LaTeX/contrib/supported/revtex4` directory.

Rotating

The *rotating* package implements three environments within which figures, tables, and captions can be rotated counterclockwise by an arbitrary number of degrees.

Note that the TrueTeX Previewer provided with *SWP* and *SW* doesn't support rotation; you must use a different DVI previewer and print driver if you want to use the rotating package to rotate figures, tables, and captions in a DVI file. The package options include many drivers. However, PDF viewers support rotation, so you can use the package to rotate figures, tables, and captions in typeset PDF files.

▶ **To rotate figures, tables, or captions**

1. Add the *rotating* package to your document.

2. Place the insertion point where you want the rotation to begin.

3. Begin the rotation environment:

   a. Enter an encapsulated TeX field.

   b. In the entry area, type \begin{*command*}{*x*} where *command* is one of the available environments:

Environment	Effect
Sideways	Print the contents of the environment turned 90 degrees counterclockwise
Turn	Print the contents of the environment turned an arbitrary number of degrees counterclockwise
Rotate	Print the contents of the environment turned an arbitrary number of degrees counterclockwise (space for the rotated results isn't necessarily created)

   and *x* is the number of degrees to be rotated. The **sideways** command doesn't require a degree argument.

   c. Choose **OK**.

4. Enter the figure, table, or caption to be rotated.

5. End the rotation environment:

   a. Enter an encapsulated TeX field.

   b. In the entry area, type \end{*command*} where *command* is the rotation environment.

   c. Choose **OK**.

The package is installed in the `TCITeX/TeX/LaTeX/contrib/supported/rotating` directory.

Scalefnt

The *scalefnt* package implements a command that scales the current font according to a specified scale factor. A factor of 2 doubles the size of the current font, like this. A factor of .5 reduces it by half, like this. You may specify any scale factor. With scalable fonts, LaTeX uses the requested font size. With bitmap font sizes, LaTeX rounds to the nearest available size.

▶ **To scale the size of the current font**

1. Add the *scalefnt* package to your document.
2. Place the insertion point where you want the font scaling to begin.
3. Enter an encapsulated TeX field.
4. Type **begingroup** and choose OK.
5. Enter an encapsulated TeX field.
6. Type **scalefont{*x*}** where *x* is the scale factor.
7. Choose OK.
8. Place the insertion point where you want the font scaling to end.
9. Type **endgroup** and choose OK.

If you don't surround the \scalefont{x} command with the encapsulated TeX fields containing the \begingroup and \endgroup commands, the font scaling will apply to the remainder of the document.

You can reverse the font scaling by using the reciprocal of the first scale factor in a second scalefont command. For example, you can begin text that is one and one half times normal size by inserting the TeX command \scalefont{1.5}. Later, you can return to the normal font size by inserting .66667, the decimal equivalent of $\frac{1}{1.5}$ (the reciprocal of the scalefont factor) in the TeX command \scalefont{.66667}.

No options are available. The package is installed in the `TCITeX/TeX/LaTeX/contrib/supported/carlisle` directory.

Sectsty

The *sectsty* package modifies the typeset appearance of section headings in articles, books, reports, and other standard documents. The package commands combine with standard LaTeX font selection commands to affect the font family, size, justification, use of rules, and numbering of division headers. Thus, you can change the text of the section headings from a sans serif font, as used in this manual, to a serif font, such as Computer Modern Roman. Using the package to change the justification for headings that are automatically indented is possible but not advisable. The *sectsty* package doesn't work with documents created with Style Editor shells.

▶ **To change the typeset appearance of section headings**

1. Add the *sectsty* package to your document.

2. From the **Typeset** menu, choose **Preamble** and click the mouse in the entry area.

3. On a new line, add the command *headingcommand*{*fontcommand*} where ***headingcommand*** is the package command indicating the section heading you want to change and ***fontcommand*** is one or more standard LaTeX font selection commands (see below).

4. Repeat step 3 for each type of section heading you want to modify.

5. Choose **OK**.

The package has no options. Available package commands include, but aren't limited to, these:

Command	Effect
\allsectionsfontv	Changes the appearance of all section headings
\partfont	Changes the appearance of all part headings and numbers
\chapterfont	Changes the appearance of all chapter headings and numbers
\sectionfont	Changes the appearance of all section headings
\subsectionfont	Changes the appearance of all subsection headings
\partnumberfont	Changes the appearance of all part numbers
\parttitlefont	Changes the appearance of all part titles only
\chapternumberfont	Changes the appearance of all chapter numbers
\chaptertitlefont	Changes the appearance of all chapter titles only

See the package documentation for a complete listing of available commands.

Standard LaTeX font selection commands include those to select a font family, such as \sffamily or \ttfamily; a font shape, such as \itshape or \textsc; a font series (width and weight), such as \bfseries or \mdseries; or a font size, such as \Large or \huge. Font selection commands also include those related to justification: \centering, \raggedright, and \raggedleft.

Thus, although the basic \section{\fontcommand} command syntax is the same, the actual commands can vary considerably depending on the extent of the modifications you want to make. An example shows the possibilities available with the package. The command \allsectionsfont{\raggedleft} right-justifies all headings in a standard LaTeX report, changing their appearance from this:

Chapter 1

This Is a Chapter

Some text goes here.

1.1 This Is a Section

Some text goes here.

1.1.1 This Is a Subsection Heading

Some text goes here.

to this:

> # Chapter 1
> # This Is a Chapter
>
> Some text goes here.
>
> ## 1.1 This Is a Section
>
> Some text goes here.
>
> ### 1.1.1 This Is a Subsection Heading
>
> Some text goes here.
>
> 3

Because *sectsty* redefines LaTeX sectioning commands, any package that requires sectioning commands won't necessarily work with the *sectsty* package. The package works for standard document classes only. The package is installed in the `TCITeX/TeX/LaTeX/contrib/supported/sectsty` directory.

Setspace

The *setspace* package replaces the *doublespace* package. It produces double and one-and-one-half line spacing based on the point size in use. The default is single spacing. With *setspace*, you can set the overall document spacing and the spacing for portions of the document. Changing spacing in the entire document doesn't affect footnotes, for which you must specify spacing individually.

In addition to the package commands described below, a linespacing option is available through the **Options and Packages** command on the **Typeset** menu.

▶ **To change the spacing for the entire document**

1. Add the *setspace* package to your document.

2. Modify the package options to select the spacing you want.

3. Choose **OK**.

▶ **To change the spacing for a portion of the document**

1. Add the *setspace* package to your document.

2. Place the insertion point where you want the spacing to change.

3. Enter a TeX field.

4. Type **singlespacing** or **onehalfspacing** or **doublespacing** and choose **OK**.

   or

   Type **setstretch**{*x*} where *x* is a number indicating the spacing you want, and then choose **OK**.

   For example, the command \setstretch{3} produces triple spacing.

5. Place the insertion point where you want to return to the original spacing.

6. Repeat steps 3–4.

Because the onehalfspace environment increases the spacing, you should avoid using it in a double-spaced document. The *setspace* package is installed in the TCITeX/TeX/LaTeX/contrib/supported/setspace directory.

Showidx

The *showidx* package prints the arguments of all index commands in the margin of the page on which they occur. The package facilitates troubleshooting and checking index entries. If you use the package to develop your index, remember to remove it before you typeset the final copy. Besides adding the package to your document, no action is required to use *showidx*. No package options are available. The package is installed in the TCITeX/TeX/LaTeX/base directory.

Showkeys

The *showkeys* package prints the keys for each label, cross-reference, page reference, citation, and bibliography item in your draft document. The package prints keys for labels and bibliography items in small boxes in the margin. Keys for cross-references, page references and citations appear where they occur, in small type elevated above the line. The package simplifies management of keys while documents are in development.

Besides adding the package to your document, no action is required to use *showkeys*. However, the package options include displaying reference and citation labels and showing labels in color. When you're ready to print the final copy of your document, choose the **final** option to suppress the action of the package or remove the package altogether. The package works with the fleqn option, $\mathcal{AMS}$-LaTeX packages, and the *varioref, natbib,* and *harvard* packages. The *showkeys* package is installed in the `TCITeX/TeX/LaTeX/required/tools` directory as part of the Standard LaTeX Tools Bundle.

Showlabels

The *showlabels* package facilitates troubleshooting of keys and markers, including the keys for automatically generated equation numbers. The package prints all keys and markers in the margin of the document at the line on which they are defined. Besides adding the package to your document, no action is required to use *showlabels*. However, you can place the labels in the inside or outside margin with the option, available through the **Options and Packages** command on the **Typeset** menu. Remember to remove the package before you typeset the final copy of your document.

The package works in two-column formats but doesn't work well with the *multicol* package. The package is installed in the `TCITeX/TeX/LaTeX/contrib/supported/showlabels` directory.

Slashed

The *slashed* package provides commands that place a slash through a character to produce the *Feynman slashed character* notation. The package works in mathematics. Some customizing of the placement of the slash is possible. Note that the package doesn't produce attractive results with some italic fonts.

You can use the package to produce Feynman characters in two ways: with TeX fields or with TeX fields in combination with negated characters. The second method modifies the program file `tcilatex.tex` to redefine the way a negated character is typeset. Because `tcilatex.tex` isn't used if you save your document with the Portable LaTeX filter, the method is incompatible with Portable LaTeX files.

▶ **To produce a Feynman slashed character with TeX fields**

1. Add the *slashed* package to your document.

2. Place the insertion point where you want the slashed character to appear.

3. Start mathematics.

4. Enter a TeX field.

5. Type **\slashed{***x***}** where *x* is the upper- or lowercase letter you want to slash.

6. Choose **OK**.

▶ **To produce a Feynman slashed character with negated characters**

1. Add the *slashed* package to your document.

2. Place the insertion point in text in your document before the first use of a slashed character.

3. Enter an encapsulated TeX field.

4. In the entry area, type **\def\NEG#1{\ensuremath{\slashed{#1}}}**.

5. Choose **OK**.

6. Place the insertion point where you want the slashed character to appear.

7. Enter the character you want to use.

8. Click [icon] ; from the **Edit** menu, choose **Properties**; or press CTRL+N.

9. Check the **Negate** box.

10. Choose **OK**.

11. Repeat steps 5–8 for any additional slashed characters you want.

No options are available. The slashed package installs in the `TCITeX/TeX/LaTeX/contrib/supported/carlisle` directory.

Subfigure

The *subfigure* package supports the use of small figures and tables within a single floating figure or table environment. The package supports the positioning, captioning, and labeling of small figures and tables and, perhaps most significantly, the inclusion of their captions in the list of figures or tables.

Several options, available through the **Options and Packages** command on the **Typeset** menu, define the placement and caption formatting of all subfigures in the document. Commands in TeX fields in the document body define specific subfigures.

The following instructions don't apply to Portable LaTeX documents, because the `FRAME` macro that defines graphics is defined in `tcilatex.tex`, which isn't included in Portable LaTeX files.

▶ **To create subfigures or subtables with the *subfigure* package**

1. Add the *subfigure* package to your document and modify the package options as necessary.

2. Place the insertion point where you want the subfigures to appear.

3. Create a floating environment:

   a. Enter an encapsulated TeX field.
   b. Type \begin{figure} to specify a floating environment.
   c. If you want a caption to apply to all the subfigures in the environment, type \caption{*title*} where *title* is the caption you want.
      This caption will appear in the list of figures for your document.
   d. If you want to create cross-references to the environment, type \label{*x*} where *x* is the key for the floating environment.
   e. Choose OK.

4. Prepare each subfigure:

   a. Open a new, completely empty document.
   b. From the File menu, choose Document Info and choose the Save Options tab.
   c. Uncheck Save Relative Graphics Paths and choose OK.
   d. From the File menu, choose Import Picture to import the graphic image you want as a subfigure.
   e. Choose Properties and then choose the Layout tab.
   f. In the Placement area, check In Line.
   g. Make any other modifications you need and then choose OK.
   h. Save the document.
   i. Open the .tex file with an ASCII editor.
   j. Find the lines that represent the figure and copy them to the clipboard.
      The lines will have an appearance something like this:

      ```
      \FRAME{dtbpF}{3.2534in}{2.2589in}{0pt}{}{}{newdoc35.wmf}{\special{language
      "Scientific Word";type "GRAPHIC";maintain-aspect-ratio TRUE;display
      "USEDEF";valid_file "F";width 3.2534in;height 2.2589in;depth
      0pt;original-width 5.3748in;original-height 3.7187in;cropleft "0";croptop
      "1";cropright "1";cropbottom "0";filename
      '../graphics/newdoc35.wmf';file-properties "XNPEU";}}
      ```

5. Return to your original *SWP* or *SW* document and reopen the encapsulated TeX field for the floating environment.

6. Create the subfigure:

   a. Place the insertion point on the line following the \begin{figure} command.
   b. Type \subfigure.
   c. If you want a caption for the subfigure, type [*title*] where *title* is the subfigure caption.
      This caption will appear in the list of figures for your document.
   d. Type {.
   e. If you want to create cross-references to the subfigure, type \label{*x*} where *x* is the key for the subfigure.
   f. Paste the lines representing the figure from the clipboard to the TeX field.
   g. Type } and choose OK.

7. Repeat steps 4–6 for each subfigure in the floating environment.

8. End the floating environment:

    a. Reopen the encapsulated TeX field for the floating environment.
    b. At the end of the entries, type **end{figure}** and choose **OK**.

9. Save and typeset compile your document.

▶ **To include subfigures in the list of figures**

1. On the Typeset toolbar, click the Front Matter button or, from the **Typeset** menu, choose **Front Matter**.

2. Place the insertion point after the **Make TOC** field.

3. Press ENTER and apply the Remove Item Tag.

4. Apply the Make LOF tag and choose **OK**.

5. From the **Typeset** menu, choose **Preamble**.

6. Click the mouse in the entry area.

7. On a new line at the end of the entries, type **setcounter{lofdepth}{2}** or **setcounter{lotdepth}{2}** and then choose **OK**.

The package works with the *caption* and *caption2* packages and should be compatible with other packages that modify or extend the float environment. You can find more information in the sample document `PackageSample-subfigure.tex` in the `SWSamples` directory of your program installation. The package is installed in the `TCITeX/TeX/LaTeX/contrib/supported/subfigure` directory.

Supertabular

The *supertabular* package defines two environments, supertabular and supertabular*, that support tabular environments longer than a single page. The package creates a separate tabular environment for each page. Therefore, the column widths for a continuing table may vary from page to page. (The *longtable* package described on page 118, avoids varying column widths for extended tables.)

The package supports a series of commands for defining the contents of the column headings, any material to be inserted at the end of each page of the environment, and a table caption. See the package documentation for complete information. The basic \supertabular command is similar to the standard LaTeX \tabular command; please refer to LaTeX sources for further information about using commands of this type. Additionally, a package option determines the extent of error information that is written to the .log file.

▶ **To create a multipage table within a supertabular environment**

1. Add the *supertabular* package to your document.

2. Enter an encapsulated TeX field.

3. Enter the commands for the entire supertabular environment, beginning with **begin{supertabular}** and ending with **end{supertabular}**.

4. Choose OK.

If you place a supertabular environment inside or on the same page as a floating element, the results are unpredictable. The *supertabular* package is installed in the `TCITeX/TeX/LaTeX/contrib/supported/supertabular` directory.

Tabularx

The *tabularx* package defines a tabular environment that generates a table with a specified width. The package automatically adjusts the widths of certain columns rather than adding space between columns. The basic command is similar to the standard LaTeX \tabular* command; please refer to the package documentation for complete instructions and to LaTeX sources for further information about using commands of this type.

▶ **To create a tabular environment of a specified width**

1. Add the *tabularx* package to your document.

2. Enter an encapsulated TeX field.

3. Type **begin{tabularx}**{*w*}{|X|X...|} where *w* is the desired width of the table and *X* marks each column for which the width is to be adjusted.

4. Type the commands for the remainder of entire supertabular environment.

5. Type **end{tabularx}**.

6. Choose OK.

The single package option involves debugging. The package is installed in the `TCITeX/TeX/LaTeX/required/tools` directory as part of the Standard LaTeX Tools Bundle.

Textcase

The *textcase* package implements commands that change the case of material in the command argument. The commands change the case of text but leave unchanged any sections of mathematics and any key names, so that references, citations, and labels still work correctly.

▶ **To change the case of specified text**

1. Add the *textcase* package to your document.

2. Enter an encapsulated TeX field.

3. Type \lowercase{text} or \uppercase{text} where *text* is the text you want changed to lowercase or uppercase.

4. Choose **OK**.

The package option addresses the use of standard text case macros. The package is installed in the `TCITeX/TeX/LaTeX/contrib/supported/carlisle` directory.

Theorem

The *theorem* package customizes theorem environments to meet the layout requirements of different journals. Theorem styles determine the appearance of theorem environments. These styles have been defined:

Style	Effect
plain	Similar to the original LaTeX definition, with the addition of parameters that specify the amount of space to skip before and after the theorem
break	Place a line break after the theorem header
marginbreak	Set the theorem number in the margin and place a line break after the theorem header
changebreak	Interchange the header number and text and place a line break after the theorem header
change	Interchange the header number and text but don't place a line break after the theorem header
margin	Set the theorem number in the margin; don't insert a line break

▶ **To customize the appearance of theorem environments**

1. Add the *theorem* package to your document.

2. From the **Typeset** menu, choose **Preamble**.

3. Click the mouse in the entry area.

4. Place the insertion point on a new line in the preamble, before the \newtheorem statement for the theorem environment you want to customize.

5. If you want the style to apply to all theorem-like objects in the document, type \theoremstyle{*style*} where *style* is the style you want.

   or

If you want the style to apply to a group of theorem-like objects in the document, type {**theoremstyle**{*style*}, where *style* is the style you want, and then place the insertion point at the end of the group and type } (closing curly brace).

6. If you want to change the font that LaTeX uses for the header of the theorem environment, type **theoremheaderfont**{*font*} where *font* is the font family you want LaTeX to use.

7. If you want to change the font that LaTeX uses for the body of the theorem environment, type **theorembodyfont**{*font*} where *font* is the font family you want LaTeX to use.

For example, to use upright text in the body of a theorem, use the command

\theorembodyfont{\upshape}.

8. Choose **OK**.

No options are available through the **Options and Packages** command on the **Typeset** menu. The *theorem* package is installed in the `TCITeX/TeX/LaTeX/required/tools` directory as part of the Standard LaTeX Tools Bundle.

Times

See PSNFSS Packages on page 127.

Tocbibind

The package automatically includes the table of contents, list of figures, list of tables, bibliography, and index in the table of contents of your document. The package is designed for use with the standard LaTeX document classes, but may cause difficulties with other document classes. The *tocbibind* package doesn't work with the version of LaTeX included with Version 3.0.

No action is needed other than adding the package to your document, but options are available from the **Options and Packages** command on the **Typeset** menu to exclude front and back matter elements from the table of contents and to use and format section headings in the table of contents. The package is installed in the `TCITeX/TeX/LaTeX/contrib/supported/tocbibind` directory.

Ulem

The *ulem* package provides several styles of simple underlines and strikethroughs by temporarily changing the behavior of the \em and \emph commands. The underlining and strikethroughs can extend across line breaks and apply to both text and mathematics.

The options available through the **Options and Packages** command on the **Typeset** menu determine how the emphasize tag is applied. The default, which you can turn off, is a single underline. As illustrated below, the package also supports double underlining, wavy underlining, a single line drawn through text, and text marked over with slashes. It supports the use of a wavy underline in place of bold text.

This package supports single and double underlining, wavy underlining, ~~a single line drawn through text~~, and ////text////marked////over////with////slashes////.

▶ **To add simple underlines**

1. Add the *ulem* package to your document.

2. Select the information you want to underline.

3. Apply the Emphasize tag.

▶ **To add varied underlines and strikethroughs**

1. Add the *ulem* package to your document.

2. Place the insertion point where you want the underline or strikethrough to begin.

3. Enter an encapsulated TeX field.

4. In the entry area, type *command*{*text*} where *command* is one of the following:

Command	Effect
\uline	Single underline
\uuline	Double underline
\uwave	Wavy underline
\sout	Line through text
\xout	Text marked over with slashes

and *text* is the information you want emphasized.

5. Choose **OK**.

In addition to using the package options, you can add underlines and strikethroughs at specific points in the document. The *ulem* package is for use with LaTeX or plain TeX. It is installed in the `TCITeX/TeX/LaTeX/contrib/other/misc` directory.

Url

The *url* package allows spacing and line breaks that result in intelligent printing of email addresses, hypertext links, and path or directory addresses. You must enter package commands in TeX fields. The address, link, path, or directory address specified in the command must not contain unbalanced braces. If it doesn't contain certain other characters (such as % or #) and doesn't end with a backslash, you can use the command in the argument to another command. Most of the path names appearing in this manual have been formatted with the aid of the *url* package.

▶ **To implement intelligent printing of hypertext links and electronic addresses**

1. Add the *url* package to your document and set the options as necessary.

2. Place the insertion point where you want the link or address to appear.

3. Enter an encapsulated TeX field.

4. In the entry area, type \url{*address*} where ***address*** is the address or link to be printed.

5. Choose OK.

In addition to package commands, the package options control spacing and line breaks. The package installs in the `TCITeX/TeX/LaTeX/contrib/other/misc` directory.

Varioref

The *varioref* package enhances page references with text that varies depending on the relative typeset location of the referenced key. The package uses standard references (using a TeX \ref command) when the command and the key occur on the same page. However, when the command and key vary by a page or more, the package inserts strings such as *on the facing page, on the preceding page, on the following page,* or *on the next page.* When the difference is greater than one page, *varioref* inserts both an enhanced reference and a standard reference. The varioref package supports *babel* so that the strings produced are customized for different languages. Also, you can customize the text strings as necessary.

▶ **To enhance page references**

1. Add the *varioref* package to your document.

2. Place the insertion point where you want the reference to appear.

3. Enter an encapsulated TeX field.

4. In the entry area, type one of the *varioref* commands:

Command	Effect
\vref{*key*}	Create an enhanced reference
\vpageref[*text*]{*key*}	Create an enhanced page reference
\vrefrange{*key*}{*key*}	Create an enhanced range of references
\vpagerefrange[*text*]{*key*}{*key*}	Create an enhanced range of page references

where ***key*** is the key you want to refer to and ***text*** is the enhancement text you want to use if the keys and references are on the same page.

5. Choose OK.

Additionally, a package option is available to aid troubleshooting. The varioref package is installed in the `TCITeX/TeX/LaTeX/required/tools` directory as part of the Standard LaTeX Tools Bundle.

Verbatim

The verbatim environment allows for the display of information exactly as it is entered at a terminal. The *verbatim* package improves that environment by handling text of arbitrary length, even an entire file, as verbatim input. The package also improves the detection of the verbatim environment's closing delimiter. You can display verbatim text with a font different from the default typewriter font.

▶ **To display verbatim text**

1. Add the *verbatim* package to your document.
2. Place the insertion point where you want the verbatim text to appear.
3. Enter an encapsulated TeX field.
4. In the entry area, type **\begin{verbatim}** and choose **OK**.
5. Type the information as you want it to appear.
6. At the end of the information, enter another encapsulated TeX field.
7. In the entry area, type **\end{verbatim}** and choose **OK**.

▶ **To import a verbatim file**

1. Add the *verbatim* package to your document.
2. Place the insertion point where you want the file to appear.
3. Enter an encapsulated TeX field.
4. In the entry area, type **\verbatiminput{***filename***}** where *filename* is the complete path name of the file to be imported, with forward slashes substituted for backslashes.
5. Choose **OK**.

The package has no options. It installs in the `TCITeX/TeX/LaTeX/required/tools` directory as part of the Standard LaTeX Tools Bundle.

Wrapfig

The *wrapfig* package allows text to be wrapped around floating objects at the side of the page, as shown here. The package provides two environments, wraptable and wrapfigure. These environments are not regular floats and may print out of sequence, but accompanying captions are correctly numbered. The package 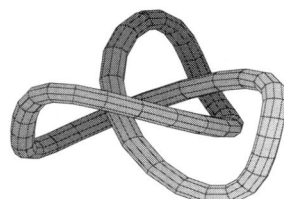 has no options available through the **Options and Packages** command on the **Typeset** menu. Find additional documentation in the `PackageSample-wrapfig.tex` file in the `SWSamples` directory of your program installation.

▶ **To wrap text around a floating figure or table**

1. Add the *wrapfig* package to your document.

2. Enter an encapsulated TeX field.

3. In the entry area, type **begin{wrapfigure}**[*w*]{*x*}[*y*]{*z*} where

    w is the number of vertical lines to be narrowed to accommodate the figure or table. We recommend you use this optional argument.

    x is the placement of the figure or table (required). Uppercase indicates *float;* lowercase indicates *exactly here*:

Placement	Effect
r or R	Right side of text
l or L	Left side of text
i or I	Inside edge, near the binding (for two-sided documents)
o or O	Outside edge, away from the binding (for two-sided documents)

    y is the amount of overhang—the distance the figure or table should extend into the margin (optional).

    z is the width of the figure or table (required). If you specify a width of zero (0pt), the package uses the actual width of the figure or table to determine the wrapping width.

4. Choose **OK**.

5. Enter the figure or table as an in-line object.

6. Enter an encapsulated TeX field.

7. In the entry area, type **end{wrapfigure}** and choose **OK**.

The package has one option available to print information in the .log file. You can find additional documentation in the SWSamples directory of your program installation. The *wrapfig* package is installed in the TCITeX/TeX/LaTeX/contrib/other/misc directory.

Xr

The *xr* package uses standard TeX \ref and \pageref commands to create cross-references and page references to labels outside the current document. You can declare as many external documents as you want. You must compile your document outside *SWP* or *SW*.

▶ **To create cross-references and page references to labels in other documents**

1. Add the *xr* package to your document.

2. Declare the external documents:

    a. From the **Typeset** menu, choose **Preamble**.
    b. Click the mouse in the entry area.

c. On a new line, type this command to declare the document for which you want to create a cross-reference: \externaldocument{*file*} where *file* is the name of another document.
   Note: Don't include the file extension.
   d. Repeat step c for as many files as necessary.
   e. Choose OK.

3. Place the insertion point where you want the cross-reference to occur.

4. Enter a TeX field.

5. In the entry area, type \ref{*label*} or \pageref{*label*} where *label* is the label in the external document.

6. Choose OK.

   No options are available. The *xr* package is installed in the `TCITeX/TeX/LaTeX/required/tools` directory as part of the Standard LaTeX Tools Bundle.

Xtab

The *xtab* package improves the page breaking capabilities of the *supertabular* package (see page 139). The package allows headings on the last page of a table to differ from those on earlier pages. Be sure to run LaTeX twice when this package is in use. A debugging option is available through the Options and Packages command on the Typeset menu.

▶ **To modify supertabular environments**

1. Add the *xtab* and *supertabular* packages to your document.

2. Enter an encapsulated TeX field.

3. Enter the commands for the entire supertabular environment, beginning with \begin{supertabular} and ending with \end{supertabular}.

4. Modify the supertabular headings on the last page as necessary by adding *xtab* \tablefirsthead and \tablelasthead commands at appropriate points in the environment. See the package documentation for a list of available commands.

5. Choose OK.

The package is installed in the `TCITeX/TeX/LaTeX/contrib/supported/xtab` directory.

Index

acronym package, 78
adding packages, 65
adding shells, 71
address for technical support, xv
afterpage package, 80
algorithm package, 81
algorithms, entering, 81
alignment of table cells, 86
alltt package, 82
American Psychological Association (APA)
    bibliography format, 85, 121
    citation format, 84
    newapa package, 121
AMS fonts
    AMSFonts package, 83
    Euler characters, 104
AMS packages, 83
AMSFonts package, 83
AMSMath package, 84
AMSSymb package, 83
answers package, 84
APA format, 84, 85, 121
apacite package, 84
apalike package, 85
apalike-plus package, 85
appendices
    for each chapter, 21, 86
    for each section, 86
    in table of contents, 19
    modifying titles, 86
    numbering scheme, 21
appendix package, 21, 86
array package, 86
arrays, *see also* tables
    aligning cells, 91
    aligning columns, 100
    colored background, 98
    delimiters around, 100
    formatting, 86, 91
    lines, 113
article.cls
    option defaults, 55
    page layout, 57

astron package, 87
author index, 21
author-date citations, 87, 88
authordate packages, 88
Avant Garde fonts, 128

babel package, 89
bibliographies
    APA format, 85, 121
    Chicago Manual of Style format, 95
    customizing, 85, 95, 112
    default name, 27, 94
    for each chapter, 93, 121
    Harvard format, 111
    improving spacing, 90
    in table of contents, 18, 142
    troubleshooting, 101, 136
    uncited items, 101
    unnumbered, 20
bibmods package, 90
BIBTeX bibliographies
    APA format, 84, 85, 121
    author-date citations, 87, 88
    Chicago Manual of Style format, 95
    citation styles, 112
    customizing, 85
    for each chapter, 93
    Harvard format, 111
blkarray package, 91
body text, 5
book.cls
    option defaults, 55
    page layout, 58
Bookman fonts, 128
boxed text, 91
boxedminipage package, 91
boxes
    color, 97
    creating, 91, 105
breakcites package, 92

caption/caption2 packages, 37, 92
captions
    customizing, 92

font attributes, 37
    formatting, 36
    rotating, 92
    spacing, 36
cells, formatting, 91
chapterbib package, 93
chapters
    appendices, 21, 86
    bibliographies for, 93, 121
characters
    AMS packages, 83
    Euler fonts, 104
    large, 104
    list of symbols, 121
    mathematical, 115
    mathematical symbols, 104
Charter fonts, 128
chbibref package, 27, 94
Chicago Manual of Style, 95
chicago package, 95
choosing shells, 61
citation labels, 101
citations
    APA format, 84
    author-date, 88
    breaking across lines, 92
    customizing, 84, 87
    for draft documents, 101
    for unnumbered bibliographies, 20
    formatting, 96, 112, 124
    ordering, 96
    troubleshooting, 101, 136
    uncited bibliography items, 101
cite package, 96
class options
    about, 54, 55
    modifying, 13, 15, 64
.clo files, 53
.cls files
    about, 53
    article.cls, 55
    book.cls, 55
    page layouts, 57
    report.cls, 56
color

in boxes, 97
in documents, 32, 97
in tables, 98
color package, 32, 97
colortbl package, 98
columns
    aligning, 100
    color, 98
    creating, 12, 121
    delimiters, 91
    floating objects, 107
    footnotes in, 110
    headers, 107
comma package, 99
conventions, xi
counters
    for numbered lists, 28
    for section numbers, 25
    for table of contents, 17
    resetting theorem counter, 33, 34
Courier fonts, 128
cross-references, 114, 136, 146
.cst files
    about, 53
    modifying, 74
customer support, xiv

damaged documents, 50
dcolumn package, 100
defaults
    bibliography name, 27, 94
    for document classes, 55
    for packages, 67
    shell, 62
delarray package, 100
delimiters in arrays, 91, 100
device independent (DVI) files, ix
document class options
    about, 55
    modifying, 13, 15, 64
document classes, 54
document language, 89
document repair, 50
document shells
    about, 53
    adding, 71
    available shells, 61
    choosing, 62

default shell, 62
from outside sources, 71
layouts, 116
saving, 70
standard LaTeX shells, 62
tailoring, 63
double columns
    creating, 12
    floating objects, 107
    footnotes in, 110
    headers, 107
    multiple columns, 121
double spacing
    creating, 3
    double-spaced documents, 101
    multiple line spacing, 135
doublespace package, 101
draft documents, 101, 135, 136
drftcite package, 101
dropped letters, 102
dropping package, 102
DVI drivers
    and color, 32, 97
    and packages, 77
    page orientation, 13
    previewing, 77
    rotation, 92, 105, 111, 118, 130
DVI files, ix

email addresses, typesetting, 143
email for technical support, xiv
Encapsulated PostScript (EPS) graphics, 40
encapsulated TeX fields, 69
endnotes package, 31, 102
endnotes, creating from footnotes, 31, 102
enumerate package, 103
EPS graphics, 40
error messages, 48
Euler fonts, 104
euler package, 104
exam questions
    answers, 84
    points for, 125
exceptional pages, 8

exercise solutions, 84
exporting shells, 70
exscale package, 104

fancybox package, 105
fancyhdr package
    page layout, 7, 8, 105
    page numbers, 11
fax number for technical support, xv
figures
    and color, 97
    captions, 36
    EPS graphics, 40
    formatting, 36, 111
    rotated, 111, 130
    spacing, 36
    within figures, 137
    wrapping text around, 39, 145
fix2col package, 107
flafter package, 108
float package, 41, 108
floating objects
    captions, 36, 92
    figures within figures, 137
    forcing output, 41, 80
    in double columns, 107
    managing, 41
    output sequence, 107
    placing, 36, 108
    titles, 36
    wrapping text around, 39, 145
font size, 5
fonts
    AMS packages, 83
    attributes for captions, 37
    changing body text size, 5
    Euler fonts, 104
    font encoding, 115
    in theorems, 35, 141
    mathematics, 104
    NFSS Mathtime fonts, 120
    NFSS Times font, 128
    PostScript fonts, 127
    sampling appearance, 109
    scaling, 131
    sizes for captions, 37
    verbatim, 145
fontsmpl package, 109

Index

footers
  customizing, 110
  eliminating space, 6, 7
  modifying, 5, 11, 105
  page numbers, 11
  specifying information, 8
  suppressing, 6
footmisc package, 110
footnotes, *see also* endnotes
  changing to endnotes, 31, 102
  for double-column text, 110
  formatting, 110
  in tables and arrays, 91
formatting
  arrays, 86, 91
  bibliographies, 85, 95, 112
  body text, 5
  captions, 36
  cells, 91
  citations, 84, 87, 96, 112, 124
  columns, 12, 121
  counters, 103
  document appearance, 53
  document class options, 54, 64
  double spacing, 3, 135
  floating objects, 41
  fonts, 109, 131
  footers, 5, 105, 110
  footnotes, 110
  front matters, 10
  graphics, 36, 111
  headers, 5, 105, 110
  indention, 27, 115
  line spacing, 3, 101, 135
  lists, 28, 103, 125
  margins, 2, 110
  page layout, 2, 54, 110, 116
  page numbers, 9, 105
  page orientation, 13, 105, 118, 126
  paragraphs, 27, 115
  section headings, 6, 23, 26, 131
  shell documents, 53
  table of contents, 17, 142
  tables, 86, 91, 98, 100, 113
  theorems, 35, 141

typesetting specifications, 53
front matter, page numbering scheme, 10
ftnright package, 110

geometry package
  changing margins, 2
  header and footer space, 6, 7, 110
  page layout, 110
  page orientation, 13, 110
  paper size, 15, 110
going native, 66
graphics
  and color, 97
  captions, 36
  EPS graphics, 40
  figures within figures, 137
  formatting, 36, 111
  rotated, 111, 130
  wrapping text around, 39, 145
graphicx package, 111

harvard package, 112
headers
  customizing, 110
  eliminating space, 6, 7
  in double columns, 107
  modifying, 5, 105
  page numbers, 11
  rules and lines, 7
  short form of headings, 6
  specifying information, 8
  suppressing, 6
headings
  adding numbers, 25
  changing appearance, 23, 131
  changing generated titles, 26
  in table of contents, 18
  location on page, 24
  numbering scheme, 26
  short form, 6
  unnumbered, 18, 25
help
  additional information, xv
  discussion forum, xv
  online Help, xiii

resources, xiii
technical support, xiv
TeX resources, viii
helvet package, 127
Helvetica fonts, 127
hhline package, 113
highlighting, xi
hyperref package, 114
hypertext links, typesetting, 114, 143
hyphenat package, 115
hyphenation, 31, 89, 115

illustrations, *see* graphics
indentfirst package, 115
indention
  paragraphs, 115
  removing from paragraphs, 27
  tables, 86
index
  author index, 21
  in table of contents, 18, 142
  makeidx package, 119
  subject index, 21
  troubleshooting index entries, 135
inputenc package, 115
Internet address for technical support, xv
interparagraph spacing, 28

keys, troubleshooting, 136

lambdabar symbol (λ), 129
landscape orientation
  changing, 13, 118, 126
  tables, 42
languages
  multiple, 89
  non-English, 89, 115
LaTeX
  about, viii
  adding counter separators, 99
  and PDFTeX, ix
  article class, 55
  book class, 55
  cautions, 1
  customizing counter style, 103

document class options, 54
document classes, 54
   packages, 77
   report class, 56
   resolving errors, 49
   Standard LaTeX shells, 62
LaTeX packages, *see* packages
latexsym package, 115
layout package, 116
letters, dropped, 102
licensing the program, xiii
line spacing
   bibliographies, 90
   changing, 3
   citations, 92
   doublespacing documents, 3, 101, 135
   multiple line spacing, 4, 135
lineno package, 116
lines
   around text, 91
   in tables, 87, 113
   over footers, 105
   under headers, 7, 105
list of acronyms, 78
list of algorithms, 81
list of figures, 16, 137, 142
list of symbols, 121
list of tables, 16, 137, 142
lists
   formatting, 103
   interrupted, 30
   numbered, 28
   spacing between items, 125
longtable package, 118
lowercase text, 140
lscape package, 118
ltxtable package, 119

MacKichan Software, contacting, xv
makeidx package, 21, 119
margins, 2, 110
markers, troubleshooting, 136
mathematical symbols, 83, 104, 115
mathematics fonts, 83, 104
mathpple package, 128
mathptm package, 128
mathptmx package, 128
mathtime package, 120
measurement units, 3
modifying packages, 67
multicol package, 12, 121
multicolumn environments, 12, 121
multiple line spacing, 3, 135

New Century Schoolbook fonts, 128
new documents, 54
New Font Selection Scheme, 115, 127
newapa package, 121
newpnts package, 125
NFSS, 115, 127
nomencl package, 21, 121
non-English documents, 89, 115
nopageno package, 123
notation, xi
numbered lists
   changing appearance, 28
   interrupted, 30
   numbering scheme, 28
   resetting, 28
   spacing, 125
numbering appendices, 21
numinsec package, 123

online Help, xiii
orientation of pages, 13, 118, 126
output drivers
   and color, 32, 97
   and packages, 77
   page orientation, 13
   previewing, 77
   rotation, 92, 105, 111, 118, 130
overcite package, 124

packages
   about, 60
   acronym, 78
   adding and removing, 61, 65
   afterpage, 80
   algorithm, 81
   alltt, 82
   AMS packages, 83
   answers, 84
   apacite, 84
   apalike, 85
   apalike-plus, 85
   appendix, 86
   array, 86
   astron, 87
   authordate1-4, 88
   babel, 89
   blkarray, 91
   boxedminipage, 91
   breakcites, 92
   caption, 92
   caption2, 92
   chapterbib, 93
   chbibref, 94
   chicago, 95
   cite, 96
   color, 97
   colortbl, 98
   comma, 99
   dcolumn, 100
   defaults, 67
   delarray, 100
   doublespace, 101
   drftcite, 101
   dropping, 102
   endnotes, 102
   enumerate, 103
   euler, 104
   exscale, 104
   fancybox, 105
   fancyhdr, 105
   fix2col, 107
   flafter, 108
   float, 108
   fontsmpl, 109
   footmisc, 110
   ftnright, 110
   geometry, 110
   going native, 66
   graphicx, 111
   harvard, 112
   hhline, 113
   hyperref, 114
   hyphenat, 115
   in use, 60
   indentfirst, 115
   inputenc, 115

Index

installed with the program, 77
latexsym, 115
layout, 116
lineno, 116
longtable, 118
lscape, 118
ltxtable, 119
makeidx, 119
mathpple, 128
mathptm, 128
mathtime, 120
modifying options, 67
multicol, 121
newapa, 121
newpnts, 125
nomencl, 121
nopageno, 123
numinsec, 123
overcite, 124
paralist, 125
pathptmx, 128
pifont, 128
points, 125
portland, 126
PSNFSS packages, 127
remreset, 129
revsymb, 129
rotating, 130
scalefnt, 131
sectsty, 131
setspace, 135
showidx, 135
showkeys, 136
showlabels, 136
slashed, 136
subfigure, 137
supertabular, 139
tabularx, 140
textcase, 140
theorem, 141
times, 128
tocbibind, 142
ulem, 142
url, 143
varioref, 144
verbatim, 145
wrapfig, 145
xr, 146
xtab, 147

page breaks, in table of contents, 17
page layout
  article.cls, 57
  book.cls, 58
  columns, 12
  customizing, 2, 110
  headers and footers, 5, 7, 105
  layout diagram, 116
  line spacing, 3, 105
  margins, 2, 105
  modifying, 2, 105
  orientation, 13, 105, 118, 126
  page numbers, 9, 11, 105
  paper size, 15
  report.cls, 59
  sebase.cls, 54
page numbers
  changing, 9, 105
  moving, 11, 105
  number style, 10, 105
  removing, 123
  resetting, 9, 105
  suppressing, 11, 105
page orientation
  changing, 13, 118, 126
  specifying, 110
page references
  adding text, 144
  troubleshooting, 136
Palatino fonts, 128
paper size, 15
paragraphs
  changing interparagraph spacing, 28
  dropped letters, 102
  indention, 115
  removing indention, 27
paralist package, 125
path names, typesetting, 143
PDF files
  and PDFTeX, viii
  color, 97
  colored tables, 98
  creating, ix
  cross-references, 114
  rotated boxes, 105
  rotated captions, 92

rotated graphics, 111
rotated graphics and tables, 130
rotated pages, 118
PDFTeX, ix
pictures, *see* graphics
pifont package, 128
points package, 125
Portable Document Format (PDF) files
  and PDFTeX, viii
  color, 97
  colored tables, 98
  creating, ix
  hypertext links, 114
  rotated boxes, 105
  rotated captions, 92
  rotated graphics, 111
  rotated graphics and tables, 130
  rotated pages, 118
portland package, 13, 42, 126
portrait orientation, 13, 126
PostScript fonts, 127
PostScript New Font Selection Scheme, 127
preamble, adding commands, 69
previewer settings, 14
previewing
  and output drivers, 77
  and typesetting, viii
printer settings, 14, 15
printing, viii, 77
programming code, typesetting, 145
PSNFSS, 127

references, *see* bibliographies
registering the program, xiii
removing packages, 65
remreset package, 129
repairing damaged documents, 50
report.cls
  option defaults, 56
  page layout, 59
revsymb package, 129
rotating package, 130
rotation
  boxes, 105

captions, 92
graphics, 111, 130
pages, 118
rows, colored background, 98
rules
  around text, 91
  in tables, 87, 113
  over footers, 105
  under headers, 7, 105

sans serif fonts, 127
saving shells, 70
scalefnt package, 131
scaling fonts, 131
sebase.cls, 54
section headings
  adding numbers, 25
  changing appearance, 23, 131
  changing generated titles, 26
  in table of contents, 18
  location on page, 24
  numbering scheme, 26
  short form, 6
  unnumbered, 18, 25
sections, appendices for, 86
sectsty package, 24, 131
selecting, xi
separators
  for LaTeX counters, 99
  for tables, 91
setspace package, 4, 135
shell documents
  about, 53
  adding, 71
  available shells, 61
  choosing, 61
  creating, 75
  default shell, 62
  from outside sources, 71
  layout, 116
  saving, 70
  standard LaTeX shells, 62
  tailoring, 63
showidx package, 135
showkeys package, 136
showlabels package, 136
single spacing, 3
slashed package, 136
spacing

changing line spacing, 3
doublespacing documents, 3, 101
in bibliographies, 90
in citations, 92
in headers and footers, 6
in lists, 125
multiple line spacing, 3, 135
single spacing, 3
standard LaTeX shells, 62
strikethroughs, 30, 142
.sty files, x, 53
subfigure package, 137
subject index, 21
supertabular package, 139
supplemental technical documents, xiv
symbols
  AMS packages, 83
  large, 104
  list of symbols, 121
  mathematical, 104, 115

table of contents
  creating page breaks in, 17
  including a bibliography, 19, 142
  including an appendix, 19
  including an index, 19
  including front and back matter, 142
  including unnumbered sections, 18
  level of headings in, 17
  short form of headings, 5
tables, see also arrays
  aligning columns, 100
  breaking at page boundaries, 147
  colored background, 98
  formatting, 86, 91
  formatting titles, 36
  landscaped, 42
  lines, 113
  long tables, 118, 139
  modifying layout, 36
  specified width, 140
  tables in tables, 137
  wrapfig package, 39
  wrapping text around, 145

tabularx package, 140
tailoring shells, 63
technical support, xiv
telephone for MacKichan Software, xv
TeX
  about, viii
  encapsulated TeX fields, 69
  entering TeX fields, 68
  measurement units, 3
  resources, viii
TeX commands, 68
text
  fitting more on page, 3
  wrapping around floating objects, 39, 145
textcase package, 140
theorem package, 35, 141
theorems
  formatting, 35, 141
  numbering schemes, 33
  resetting counters, 34
  theorem environments, 33
Times fonts, 128
times package, 128
title pages, 16
titles
  for tables, 37
  for appendices, 86
tocbibind package, 142
toll-free number for technical support, xv
troubleshooting
  bibliographies, 101, 136
  citations, 101
  cross-references, 136
  damaged documents, 50
  index entries, 135
  keys, 136
  LaTeX errors, 48
  markers, 136
  online Help, xiii
  technical support, xiv
TrueTeX, viii
typesetting specifications
  about, 53
  adding, 71
  directory assignment, 72
  installing, 72
  modifying, 42

ulem package, 30, 142
underlines, 30, 142
unnumbered bibliographies, 20
unnumbered sections, 18, 25
uppercase text, 140
url package, 143
URLs, typesetting, 143

Utopia fonts, 128

varioref package, 144
verbatim information, 82, 145
verbatim package, 145

website for technical support, xiv, xv

wrapfig package, 39, 145

xr package, 146
xtab package, 147

Zapf Chancery fonts, 128
Zapf Dingbats font, 128

Software

 ### Scientific WorkPlace®

Scientific WorkPlace makes writing, sharing, and doing mathematics easier than you ever imagined possible. This scientific word processor increases your productivity because it is easy to learn and use. You can compose and edit your documents directly on the screen, without being forced to think in a programming language. With a simple click of a button, you can typeset your document in LaTeX. You can also compute and plot solutions with the included computer algebra system. With *Scientific WorkPlace*, both professional and support staff can produce stunning results quickly and easily, without knowing TeX™, LaTeX, or computer algebra syntax. ***Contact us for a free 30-day trial version.***

 ### Scientific Word®

The Gold Standard for mathematical publishing since 1992, *Scientific Word* makes writing and sharing scientific documents straightforward and easy. With over 100 LaTeX styles included, *Scientific Word* ensures your documents will be beautiful. This means you can concentrate on the content, not the style. It has been estimated that support staff using *Scientific Word* experiences a doubling or tripling of productivity over the use of straight LaTeX. Best of all, MacKichan Software provides free, prompt, and knowledgeable technical support. ***Contact us for a free 30-day trial version.***

 ### Scientific Notebook®

Scientific Notebook makes word processing and doing mathematics easy. With this complete word processor, you can enter text and mathematics quickly, without having to use an inefficient equation editor. The built-in computer algebra system lets you solve and plot equations without having to learn a special syntax. After creating your scientific documents and exams in *Scientific Notebook*, you can publish them in print and on the World Wide Web. ***Contact us for a free 30-day trial version.***

 ### MuPAD® Pro

MuPAD Pro is a modern, full-featured computer algebra system in an integrated and open environment for symbolic and numeric computing. Its domains and categories are like object-oriented classes that allow overriding and overloading methods and operators, inheritance, and generic algorithms. The *MuPAD* language has a Pascal-like syntax and allows imperative, functional, and object-oriented programming. A comfortable notebook interface includes a graphics tool for visualization, an integrated source-level debugger, a profiler, and hypertext help. ***Contact us for a free 30-day trial version.***

TO ORDER: Visit our webstore, fax, email, or phone us.
Website: www.mackichan.com ♦ Fax: 360-394-6039 ♦ Email: info@mackichan.com ♦ Toll-free: 877-724-9673
19307 8th Avenue NE ♦ Suite C ♦ Poulsbo, WA 98370

Scientific WorkPlace®, Scientific Word®, and Scientific Notebook® are registered trademarks of MacKichan Software, Inc.
MuPAD® is a registered trademark of SciFace GmbH.
TeX™ is a trademark of The American Mathematical Society.

Additional Software

MathTalk™/Scientific Notebook®
MathTalk/Scientific Notebook, created by Metroplex Voice Computing, provides voice input for *Scientific Notebook*. With this program, you can enter even the most complex mathematics using voice commands. You can use it in conjunction with the keyboard and mouse to speed the entry of text and mathematics, or to completely replace the keyboard and mouse. *MathTalk/Scientific Notebook* requires *Dragon NaturallySpeaking*®, which is not included. Visit www.mathtalk.com for more information.

Duxbury Braille Translator
Duxbury Systems leads the world in software for braille. Their *Duxbury Braille Translator* converts *Scientific Notebook* files into braille. Create your print or large print math, save, then open your *Scientific Notebook* file and go to braille. Used with *MathTalk/Scientific Notebook*, the *Duxbury Braille Translator* provides dramatic new power to visually impaired students and professionals. Visit www.duxburysystems.com for more information.

Books

Doing Mathematics with Scientific WorkPlace® and Scientific Notebook®, Version 5
By Darel W. Hardy and Carol L. Walker
  Doing Mathematics with Scientific WorkPlace and Scientific Notebook describes how to use the built-in computer algebra system to do a wide range of mathematics, without having to deal directly with the computer algebra syntax.

Creating Documents with Scientific WorkPlace® and Scientific Word®, Version 5
By Susan Bagby
  Creating Documents with Scientific WorkPlace and Scientific Word gives you an overview of the process of creating beautiful documents using these two powerful software programs. It covers basic editing and entering of mathematical expressions as well as tables, graphics, lists, indexes, cross-references, tables of contents, and large document management. If you are using *Scientific WorkPlace* or *Scientific Word* to prepare documents for publication, this book is highly recommended.

(Continued)

TO ORDER: Visit our webstore, fax, email, or phone us.
Website: www.mackichan.com ♦ Fax: 360-394-6039 ♦ Email: info@mackichan.com ♦ Toll-free: 877-724-9673
19307 8th Avenue NE ♦ Suite C ♦ Poulsbo, WA 98370

MathTalk™ is a trademark of Metroplex Voice Computing.
Dragon NaturallySpeaking® is a registered trademark of Dragon Systems, Inc.

Books (cont.)

Typesetting Documents in Scientific WorkPlace® and Scientific Word®, Second Edition
By Susan Bagby and George Pearson

Typesetting Documents in ScientificWorkPlace and Scientific Word is an aid to choosing and customizing the typeset appearance of your documents. The manual explains how to work from within these two programs to tailor the typeset appearance of a document; how to choose and add document shells; and how document shells use LaTeX document classes, class options, and packages. The manual also documents many of the LaTeX packages that accomplish specific formatting tasks.

A Gallery of Document Shells for Scientific WorkPlace® and Scientific Word®, Version 5
By Susan Bagby and George Pearson (In PDF format on the program CD-ROM)

A Gallery of Document Shells for Scientific WorkPlace and Scientific Word helps you choose document shells that are appropriate for your typesetting purposes. It illustrates and briefly describes the characteristics of almost 200 shells provided with these two programs.

Doing Calculus with Scientific Notebook®
By Darel W. Hardy and Carol L. Walker

Take the mystery out of doing calculus with this must-have companion to the *Scientific Notebook* software. This book provides activities to complete with *Scientific Notebook* that will help develop a clearer understanding of calculus. Think of *Scientific Notebook* as a laboratory for mathematical experimentation and *Doing Calculus with Scientific Notebook* as a lab manual of experiments to perform.

An Interactive Introduction to Mathematical Analysis
By Jonathan Lewin (Cambridge University Press)

This book is a sequel to *An Introduction to Mathematical Analysis.* It includes an on-screen hypertext version that you can read with *Scientific Notebook*. This on-screen version contains alternative approaches to material, more fully explained forms of proofs of theorems, sound movie versions of proofs of theorems, interactive exploration of mathematical concepts using the computing features of *Scientific Notebook*, automatic links to the author's website for solutions to exercises, and more. It can be ordered at any bookstore.

(Continued)

TO ORDER: Visit our webstore, fax, email, or phone us.
Website: www.mackichan.com • Fax: 360-394-6039 • Email: info@mackichan.com • Toll-free: 877-724-9673
19307 8th Avenue NE • Suite C • Poulsbo, WA 98370

Books (cont.)

Precalculus with Scientific Notebook®
By Jonathan Lewin (Kendall Hunt Publishing Company)

This book contains a standard printed version and an on-screen hypertext version designed for interactive reading with *Scientific Notebook*. The on-screen version includes links to solutions to exercises that reside on the author's website. It can be ordered at any bookstore.

Exploring Mathematics with Scientific Notebook®
By Wei-Chi Yang and Jonathan Lewin (Springer-Verlag)

This book is supplied both in printed form and in an on-screen hypertext version for interactive reading with *Scientific Notebook*. It contains a sequence of modules from a variety of mathematical areas and, in each, demonstrates how the editing, Internet, and computing features of *Scientific Notebook* can be combined to deepen the reader's understanding of mathematical concepts. It can be ordered at any bookstore.

MuPAD® Pro Computing Essentials
By Miroslav Majewski (Springer-Verlag)

Intended for teachers of mathematics and their students, *MuPAD Pro Computing Essentials* presents basic information about doing mathematics with *MuPAD Pro*. It includes basic instructions useful in various areas of mathematics that are facilitated by *MuPAD Pro*. Chapters 1 through 7 focus on the basics of using *MuPAD Pro*: Syntax, programming control structures, procedures, libraries, and graphics. Chapters 8 through 13 focus on applications in geometry, algebra, logic, set theory, calculus, and linear algebra. Each chapter includes examples and programming exercises.

TO ORDER: Visit our webstore, fax, email, or phone us.
Website: www.mackichan.com ♦ Fax: 360-394-6039 ♦ Email: info@mackichan.com ♦ Toll-free: 877-724-9673
19307 8th Avenue NE ♦ Suite C ♦ Poulsbo, WA 98370

Workshops

Scientific WorkPlace®, Scientific Word®, and Scientific Notebook®
By Professor Jonathan Lewin, Ph.D.

Seminars, workshops, and training sessions by Jonathan Lewin are available at professional conferences and, by arrangement, at individual campuses of high schools, colleges, and universities.

Scientific Notebook: This presentation introduces the editing, Internet and computing features of *Scientific Notebook* documents. It also covers the publication of mathematical material on websites and how *Scientific Notebook* can be used as an electronic whiteboard in the classroom.

Scientific WorkPlace and Scientific Word: This presentation introduces the editing and typesetting features of *Scientific WorkPlace* and *Scientific Word*. Participants will be trained in the production of documents that will be printed as professional quality hard copy or submitted to an editor for publication.

Each participant will be given a CD containing sound movies that review the material covered in the workshops. Contact Jonathan Lewin for details at *lewins@mindspring.com* or 770-973-5931.

Scientific WorkPlace® and Scientific Notebook®
By Professor Bill Pletsch, Ph.D.

Bill Pletsch has been using computer algebra systems since the early 1980s. He is available for workshops and training sessions on the use of *Scientific WorkPlace* and *Scientific Notebook* as a research and teaching aid.

The workshops and training sessions begin with a demonstration of the capabilities of *Scientific WorkPlace/Notebook*. The introductory demonstration is followed by a thorough nuts and bolts hands-on session on the basics. Participants learn how to compute numerically and symbolically, graph, and manipulate data. Also included are word-processing, Internet techniques, and utilizing other resources. More advanced topics include the use of *Scientific WorkPlace/Notebook* in the classroom and in the preparation of computer classroom lectures and demonstrations.

An overview of computer algebra systems and why every teacher of mathematics should own a copy of *Scientific Notebook* will accompany the presentation. Included in the workshops will be a CD on the electronic delivery of mathematics instruction. No prior computer experience is required. Contact Bill Pletsch for details at *bpletsch@tvi.edu* or 505-224-3672.

(Continued)